Advanced Information and Knowledge Processing

Lipo Wang · Xiuju Fu

Data Mining with Computational Intelligence

With 72 Figures and 65 Tables

Lipo Wang
Nanyang Technological University
School of Electrical and Electronical Engineering
Block S1, Nanyang Avenue,
639798 Singapore, Singapore
elpwang@ntu.edu.sg

Xiuju Fu
Institute of High Performance Computing,
Software and Computing, Science Park 2,
The Capricorn
Science Park Road 01-01
117528 Singapore, Singapore
fuxj@pmail.ntu.edu.sg

Series Editors
Xindong Wu
Lakhmi Jain

ISBN-13 978-3-642-06387-9 e-ISBN-13 978-3-540-28803-9

ACM Computing Classification (1998): H.2.8., I.2

Springer is a part of Springer Science+Business Media
springeronline.com

Printed in Germany

Cover design: KünkelLopka, Heidelberg

Printed on acid-free paper 45/3142/YL - 5 4 3 2 1 0

Preface

Nowadays data accumulate at an alarming speed in various storage devices, and so does valuable information. However, it is difficult to understand information hidden in data without the aid of data analysis techniques, which has provoked extensive interest in developing a field separate from machine learning. This new field is data mining.

Data mining has successfully provided solutions for finding information from data in bioinformatics, pharmaceuticals, banking, retail, sports and entertainment, etc. It has been one of the fastest growing fields in the computer industry. Many important problems in science and industry have been addressed by data mining methods, such as neural networks, fuzzy logic, decision trees, genetic algorithms, and statistical methods.

This book systematically presents how to utilize fuzzy neural networks, multi-layer perceptron (MLP) neural networks, radial basis function (RBF) neural networks, genetic algorithms (GAs), and support vector machines (SVMs) in data mining tasks. Fuzzy logic mimics the imprecise way of reasoning in natural languages and is capable of tolerating uncertainty and vagueness. The MLP is perhaps the most popular type of neural network used today. The RBF neural network has been attracting great interest because of its locally tuned response in RBF neurons like biological neurons and its global approximation capability. This book demonstrates the power of GAs in feature selection and rule extraction. SVMs are well known for their excellent accuracy and generalization abilities.

We will describe data mining systems which are composed of data preprocessing, knowledge-discovery models, and a data-concept description. This monograph will enable both new and experienced data miners to improve their practices at every step of data mining model design and implementation.

Specifically, the book will describe the state of the art of the following topics, including both work carried out by the authors themselves and by other researchers:

- Data mining tools, i.e., neural networks, support vector machines, and genetic algorithms with application to data mining tasks.
- Data mining tasks including data dimensionality reduction, classification, and rule extraction.

Lipo Wang wishes to sincerely thank his students, especially Feng Chu, Yakov Frayman, Guosheng Jin, Kok Keong Teo, and Wei Xie, for the great pleasure of collaboration, and for carrying out research and contributing to this book. Thanks are due to Professors Zhiping Lin, Kai-Ming Ting, Chunru Wan, Ron (Zhengrong) Yang, Xin Yao, and Jacek M. Zurada for many helpful discussions and for the opportunities to work together. Xiuju Fu wishes to express gratitude to Dr. Gih Guang Hung, Liping Goh, Professors Chongjin Ong and S. Sathiya Keerthi for their discussions and supports in the research work. We also express our appreciation for the support and encouragement from Professor L.C. Jain and Springer Editor Ralf Gerstner.

Singapore,
May 2005

Lipo Wang
Xiuju Fu

Contents

1 Introduction

This book is concerned with the challenge of mining knowledge from data. The world is full of data. Some of the oldest written records on clay tablets are dated back to 4000 BC. With the creation of paper, data had been stored in myriads of books and documents. Today, with increasing use of computers, tremendous volumes of data have filled hard disks as digitized information. In the presence of the huge amount of data, the challenge is how to truly understand, integrate, and apply various methods to discover and utilize knowledge from data. To predict future trends and to make better decisions in science, industry, and markets, people are starved for discovery of knowledge from this morass of data.

Though 'data mining' is a new term proposed in recent decades, the tasks of data mining, such as classification and clustering, have existed for a much longer time. With the objective to discover unknown patterns from data, methodologies of data mining are derived from machine learning, artificial intelligence, and statistics, etc. Data mining techniques have begun to serve fields outside of computer science and artificial intelligence, such as the business world and factory assembly lines. The capability of data mining has been proven in improving marketing campaigns, detecting fraud, predicting diseases based on medical records, etc.

This book introduces fuzzy neural networks (FNNs), multi-layer perceptron neural networks (MLPs), radial basis function (RBF) neural networks, genetic algorithms (GAs), and support vector machines (SVMs) for data mining. We will focus on three main data mining tasks: data dimensionality reduction (DDR), classification, and rule extraction. For more data mining topics, readers may consult other data mining text books, e.g., [129][130][346].

A data mining system usually enables one to collect, store, access, process, and ultimately describe and visualize data sets. Different aspects of data mining can be explored independently. Data collection and storage are sometimes not included in data mining tasks, though they are important for data mining. Redundant or irrelevant information exists in data sets, and inconsistent formats of collected data sets may disturb the processes of data mining, even mislead search directions, and degrade results of data mining. This happens because data collectors and data miners are usually not from the same group, i.e., in most cases, data are not originally prepared for the purpose of data mining. Data warehouse is increasingly adopted as an efficient way to store metadata. We will not discuss data collection and storage in this book.

1.1 Data Mining Tasks

There are different ways of categorizing data mining tasks. Here we adopt the categorization which captures the processes of a data mining activity, i.e., data preprocessing, data mining modelling, and knowledge description. Data preprocessing usually includes noise elimination, feature selection, data partition, data transformation, data integration, and missing data processing, etc. This book introduces data dimensionality reduction, which is a common technique in data preprocessing. fuzzy neural networks, multi-layer neural networks, RBF neural networks, and support vector machines (SVMs) are introduced for classification and prediction. And linguistic rule extraction techniques for decoding knowledge embedded in classifiers are presented.

1.1.1 Data Dimensionality Reduction

Data dimensionality reduction (DDR) can reduce the dimensionality of the hypothesis search space, reduce data collection and storage costs, enhance data mining performance, and simplify data mining results. Attributes or features are variables of data samples and we consider the two terms interchangeable in this book.

One category of DDR is feature extraction, where new features are derived from the original features in order to increase computational efficiency and classification accuracy. Feature extraction techniques often involve non-linear transformation [60][289]. Sharma *et al.* [289] transformed features non-linearly using a neural network which is discriminatively trained on the phonetically labelled training data. Coggins [60] had explored various non-linear transformation methods, such as folding, gauge coordinate transformation, and non-linear diffusion, for feature extraction. Linear discriminant analysis (LDA) [27][168][198] and principal components analysis (PCA) [49][166] are two popular techniques for feature extraction. Non-linear transformation methods are good in approximation and robust for dealing with practical non-linear problems. However, non-linear transformation methods can produce unexpected and undesirable side effects in data. Non-linear methods are often not invertible, and knowledge learned by applying a non-linear transformation method in one feature space might not be transferable to the next feature space. Feature extraction creates new features, whose meanings are difficult to interpret.

The other category of DDR is feature selection. Given a set of original features, feature selection techniques select a feature subset that performs the best for induction systems, such as a classification system. Searching for the optimal subset of features is usually difficult, and many problems of feature selection have been shown to be NP-hard [21]. However, feature selection techniques are widely explored because of the easy interpretability of the features selected from the original feature set compared to new features transformed from the original feature set. Lots of applications, including document classification, data mining tasks, object recognition, and image processing, require aid from feature selection for data preprocessing.

Many feature selection methods have been proposed in the literature. A number of feature selection methods include two parts: (1) a ranking criterion for ranking the importance of each feature or subsets of features, (2) a search algorithm, for example backward or forward search. Search methods in which features are iteratively added ('bottom-up') or removed ('top-down') until some termination criterion is met are referred to as *sequential* methods. For instance, sequential forward selection (SFS) [345] and sequential backward selection (SBS) [208] are typical *sequential* feature selection algorithms. Assume that d is the number of features to be selected, and n is the number of original features. SFS is a bottom-up approach where one feature which satisfies some criterion function is added to the current feature subset at a time until the number of features reaches d. SBS is a top-down approach where features are removed from the entire feature set one by one until $D-d$ features have been deleted. In both the SFS algorithm and the SBS algorithm, the number of feature subsets that have to be inspected is $n+(n-1)+(n-2)+\cdots+(n-d+1)$. However, the computational burden of SBS is higher than SFS, since the dimensionality of inspected feature subsets in SBS is greater than or equal to d. For example, in SBS, all feature subsets with dimension $n-1$ are inspected first. The dimensionality of inspected feature subsets is at most equal to d in SFS.

Many feature selection methods have been developed based on traditional SBS and SFS methods. Different criterion functions including or excluding a subset of features to the selected feature set are explored. By ranking each feature's importance level in separating classes, only n feature subsets are inspected for selecting the final feature subset. Compared to evaluating all feature combinations, ranking individual feature importance can reduce computational cost, though better feature combinations might be missed in this kind of approach. When computational cost is too heavy to stand, feature selection based on ranking individual feature importance is a preference.

Based on an entropy attribute ranking criterion, Dash *et al.* [71] removed attributes from the original feature set one by one. Thus only n feature subsets have to be inspected in order to select a feature subset, which leads to a high classification accuracy. And, there is no need to determine the number of features selected in advance. However, the class label information is not utilized in Dash *et al.*'s method. The entropy measure was used in [71] for ranking attribute importance. The class label information is critical for detecting irrelevant or redundant attributes. It motivates us to utilize the class label information for feature selection, which may lead to better feature selection results, i.e., smaller feature subsets with higher classification accuracy.

Genetic algorithms (GAs) are used widely in feature selection [44][322][351]. In a GA feature selection method, a feature subset is represented by a binary string with length n. A zero or one in position i indicates the absence or presence of feature i in the feature subset. In the literature, most feature selection algorithms select a general feature subset (class-independent features) [44][123][322] for all classes. Actually, a feature may have different discriminatory capability for distinguishing different classes from other classes. For discriminating patterns of a certain class from other patterns, a multi-class data set can be considered as a two-class data set, in which all the other classes are treated as one class against the current processed class. For example, there is a data set containing the information of ostriches, parrots, and ducks. The information of the three kinds of birds includes weight, feather color (colorful or not), shape of mouth, swimming capability (whether it can swim or not), flying capability (whether it can fly or not), etc. According to the characteristics of each bird, the feature 'weight' is sufficient for separating ostriches from the other birds, the feature 'feather color' can be used to distinguish parrots from the other birds, and the feature 'swimming capability' can separate ducks from the other birds.

Thus, it is desirable to obtain individual feature subsets for the three kinds of birds by class-dependent feature selection, which separates each one from others better than using a general feature subset. The individual characteristics of each class can be highlighted by class-dependent features. Class-dependent feature selection can also facilitate rule extraction, since lower dimensionality leads to more compact rules.

1.1.2 Classification and Clustering

Classification and clustering are two data mining tasks with close relationships. A class is a set of data samples with some similarity or relationship and all samples in this class are assigned the same class label to distinguish them from samples in other classes. A cluster is a collection of objects which are similar locally. Clusters are usually generated in order to further classify objects into relatively larger and meaningful categories.

Given a data set with class labels, data analysts build classifiers as predictors for future unknown objects. A classification model is formed first based on available data. Future trends are predicted using the learned model. For example, in banks, individuals' personal information and historical credit records are collected to build a model which can be used to classify new credit applicants into categories of low, medium, or high credit risks. In other cases, with only personal information of potential customers, for example, age, education levels, and range of salary, data miners employ clustering techniques to group the clusters according to some similarities and further label the customers into low, medium, or high levels for later targeted sales.

In general, clustering can be employed for dealing with data without class labels. Some classification methods cluster data into small groups first before proceeding to classification, e.g. in the RBF neural network. This will be further discussed in Chap. 4.

1.1.3 Rule Extraction

Rule extraction [28][150][154][200] seeks to present data in such a way that interpretations are actionable and decisions can be made based on the knowledge gained from the data. For data mining clients, they expect a simple explanation of why there are certain classification results: what is going on in a high-dimensional database, and which feature affects data mining results significantly, etc. For example, a succinct description of a market behavior is useful for making decisions in investment. A classifier learns from training data and stores learned knowledge into the classifier parameters, such as the weights of a neural network classifier. However, it is difficult to interpret the knowledge in an understandable format by the classifier parameters. Hence, it is desirable to extract IF–THEN rules to represent valuable information in data.

Rule extraction can be categorized into two major types. One is concerned with the relationship between input attributes and output class labels in labelled data sets. The other is association rule mining, which extracts relationships between attributes in data sets which may not have class labels. Association rule extraction techniques are usually used to discover relationships between items in transaction data. An association rule is expressed as '$X \Rightarrow Z$', where X and Z are two sets of items. '$X \Rightarrow Z$' represents that if a transaction $T \in D$ contains X, then the transaction also contains Z, where D is the transaction data set. A confidence parameter, which is the conditional probability $p(Z \in T \mid X \in T)$ [137], is used to evaluate the rule accuracy. The association rule mining can be applied for analyzing supermarket transactions. For example, 'A customer who buys butter will also buy bread with a certain probability'. Thus, the two associated items can be arranged in close proximity to improve sales according to this discovered association rule. In the rule extraction part of this book, we focus on the first type of rule extraction, i.e., rule extraction based on classification models. Usually, association rule extraction can be treated as the first category of rule extraction, which is based on classification. For example, if an association rule task is to inspect what items are apt to be bought together with a particular item set X, the item set X can be used as class labels. The other items in a transaction T are treated as attributes. If X occurs in T, the class label is 1, otherwise it is labelled 0. Then, we could discover the items associated with the occurrence of X, and also the non-occurrence of X. The association rules can be equally extracted based on classification. The classification accuracy can be considered as the rule confidence.

RBF neural networks are functionally equivalent to fuzzy inference systems under some restrictions [160]. Each hidden neuron could be considered as a fuzzy rule. In addition, fuzzy rules could be obtained by combining fuzzy logic with our crisp rule extraction system. In Chap. 3, fuzzy rules are presented. For crisp rules, there are three kinds of rule decision boundaries found in the literature [150][154][200][214]: hyper-plane, hyper-ellipse, and hyper-rectangular. Compared to the other two rule decision boundaries, a hyper-rectangular decision boundary is simpler and easier to understand. Take a simple example; when judging whether a patient gets a high fever, his body temperature is measured and a given temperature range is preferred to a complex function of the body temperature. Rules with a hyper-rectangular decision boundary are more understandable for data mining clients. In the RBF neural network classifier, the input data space is separated into hyper-ellipses, which facilitates the extraction of rules with hyper-rectangular decision boundaries. We also describe crisp rules in Chap. 7 and Chap. 10 of this book.

1.2 Computational Intelligence Methods for Data Mining

1.2.1 Multi-layer Perceptron Neural Networks

Neural network classifiers are very important tools for data mining. Neural interconnections in the brain are abstracted and implemented on digital computers as neural network models. New applications and new architectures of neural networks (NNs) are being used and further investigated in companies and research institutes for controlling costs and deriving revenue in the market. The resurgence of interest in neural networks has been fuelled by the success in theory and applications.

A typical multi-layer perceptron (MLP) neural network shown in Fig. 1.1 is most popular in classification. A hidden layer is required for MLPs to classify linearly inseparable data sets. A hidden neuron in the hidden layer is shown in Fig. 1.2.

The jth output of a feedforward MLP neural network is:

$$y_j = f(\sum_{i=1}^{K} W_{ij}^{(2)} \phi_i(\mathbf{x}) + b_j^{(2)}), \tag{1.1}$$

where $W_{ij}^{(2)}$ is the weight connecting hidden neuron i with output neuron j. K is the number of hidden neurons. $b_j^{(2)}$ is the bias of output neuron j. $\phi_i(\mathbf{x})$ is the output of hidden neuron i. $\mathbf{x}$ is the input vector.

$$\phi_i(\mathbf{x}) = f(\mathbf{W}_i^{(1)} \cdot \mathbf{x} + b_i^{(1)}), \tag{1.2}$$

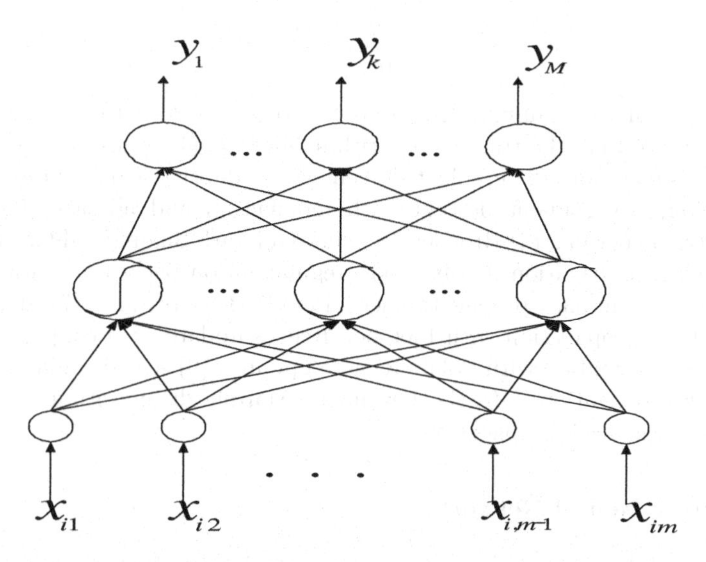

Fig. 1.1. A two-layer MLP neural network with a hidden layer and an output layer. The input nodes do not carry out any processing.

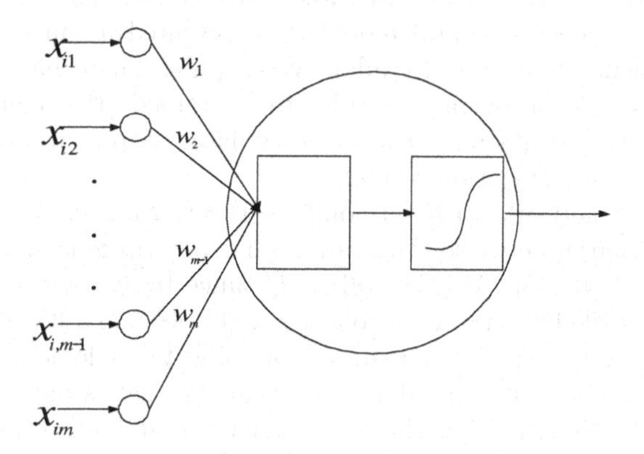

Fig. 1.2. A hidden neuron of the MLP.

where $\mathbf{W}_i^{(1)}$ is the weight vector connecting the input vector with hidden neuron i. $b_i^{(1)}$ is the bias of hidden neuron i.

A common activation function f is a sigmoid function. The most common of the sigmoid functions is the logistic function:

$$f(z) = \frac{1}{1 + \mathrm{e}^{-\beta z}}. \tag{1.3}$$

where β is the gain.

Another sigmoid function often used in MLP neural networks is the hyperbolic tangent function that takes on values between -1 and 1:

$$f(z) = \frac{e^{\beta z} - e^{-\beta z}}{e^{\beta z} + e^{-\beta z}}, \tag{1.4}$$

There are many training algorithms for MLP neural networks. As summarized in [63][133], the training algorithms include: (1) gradient descent error back-propagation, (2) gradient descent with adaptive learning rate back-propagation, (3) gradient descent with momentum and adaptive learning rate back-propagation, (4) Broyden-Fletcher-Goldfarb-Shanno (BFGS) quasi-Newton back-propagation, (5) bayesian regularization back-propagation, (6) conjugate gradient back-propagation with Powell–Beale restarts, (7) conjugate gradient back-propagation with Fletcher–Reeves updates, (8) conjugate gradient back-propagation with Polak–Ribiere updates, (9) scaled conjugate gradient back-propagation, (10) the Levenberg–Marquardt algorithm, and (11) one–step secant back-propagation.

1.2.2 Fuzzy Neural Networks

Symbolic techniques and crisp (non-fuzzy) neural networks have been widely used for data mining. Symbolic models are represented as either sets of 'IF–THEN' rules or decision trees generated through symbolic inductive algorithms [30][251]. A crisp neural model is represented as an architecture of threshold elements connected by adaptive weights. There have been extensive research results on extracting rules from trained crisp neural networks [110][116][200][297][313][356]. For most noisy data, crisp neural networks lead to more accurate classification results.

Fuzzy neural networks (FNNs) combine the learning and computational power of crisp neural networks with human-like descriptions and reasoning of fuzzy systems [174][218][235][268][336][338]. Since fuzzy logic has an affinity with human knowledge representation, it should become a key component of data mining systems. A clear advantage of using fuzzy logic is that we can express knowledge about a database in a manner that is natural for people to comprehend. Recently, there has been much research attention devoted to rule generation using various FNNs. Rather than attempting an exhaustive literature survey in this area, we will concentrate below on some work directly related to ours, and refer readers to a recent review by Mitra and Hayashi [218] for more references.

In the literature, crisp neural networks often have a fixed architecture, i.e., a predetermined number of layers with predetermined numbers of neurons. The weights are usually initialized to small random values. Knowledge-based networks [109][314] use crude domain knowledge to generate the initial network architecture. This helps in reducing the search space and time required for the network to find an optimal solution. There have also been mechanisms to generate crisp neural networks from scratch, i.e., initially there are no neurons or weights, which are generated and then refined during training. For example, Mezard and Nadal's tiling algorithm [216], Fahlman and Lebiere's cascade correlation [88], and Giles *et al.*'s constructive learning of recurrent networks [118] are very useful.

For FNNs, it is also desirable to shift from the traditional fixed architecture design methodology [143][151][171] to self-generating approaches. Higgins and Goodman [135] proposed an algorithm to create a FNN according to input data. New membership functions are added at the point of maximum error on an as-needed basis, which will be adopted in this book. They then used an information-theoretic approach to simplify the rules. In contrast, we will combine rules using a computationally more efficient approach, i.e., a fuzzy similarity measure.

Juang and Lin [165] also proposed a self-constructing FNN with online learning. New membership functions are added based on input–output space partitioning using a self-organizing clustering algorithm. This membership creation mechanism is not directly aimed at minimizing the output error as in Higgins and Goodman [135]. A back-propagation-type learning procedure was used to train network parameters. There were no rule combination, rule pruning, or eliminations of irrelevant inputs.

Wang and Langari [335] and Cai and Kwan [41] used self-organizing clustering approaches [267] to partition the input/output space, in order to determine the number of rules and their membership functions in a FNN through batch training. A back-propagation-type error-minimizing algorithm is often used to train network parameters in various FNNs with batch training [160], [151].

Liu and Li [197] applied back-propagation and conjugate gradient methods for the learning of a three-layer regular feedforward FNN [37]. They developed a theory for differentiating the input–output relationship of the regular FNN and approximately realized a family of fuzzy inference rules and some given fuzzy functions.

Frayman and Wang [95][96] proposed a FNN based on the Higgins-Goodman model [135]. This FNN has been successfully applied to a variety of data mining [97] and control problems [94][98][99]. We will describe this FNN in detail later in this book.

1.2.3 RBF Neural Networks

The RBF neural network [91][219] is widely used for function approximation, interpolation, density estimation, classification, etc. For detailed theory and applications of other types of neural networks, readers may consult various textbooks on neural networks, e.g., [133][339].

RBF neural networks were first proposed in [33][245]. RBF neural networks [22] are a special class of neural networks in which the activation of a hidden neuron (hidden unit) is determined by the *distance* between the input vector and a prototype vector. Prototype vectors refer to centers of clusters obtained during RBF training. Usually, three kinds of distance metrics can be used in

RBF neural networks, such as Euclidean, Manhattan, and Mahalanobis distances. Euclidean distance is used in this book. In comparison, the activation of an MLP neuron is determined by a dot-product between the input pattern and the weight vector of the neuron. The dot-product is equivalent to the Euclidean distance only when the weight vector and all input vectors are normalized, which is not the case in most applications.

Usually, the RBF neural network consists of three layers, i.e., the input layer, the hidden layer with Gaussian activation functions, and the output layer. The architecture of the RBF neural network is shown in Fig. 1.3. The RBF neural network provides a function $Y : R^n \rightarrow R^M$, which maps n-dimensional input patterns to M-dimensional outputs ($\{(X_i, Y_i) \in R^n \times R^M, i = 1, 2, ..., N\}$). Assume that there are M classes in the data set. The mth output of the network is as follows:

$$y_m(X) = \sum_{j=1}^{K} w_{mj} \o_j(X) + w_{m0} b_m. \tag{1.5}$$

Here X is the n-dimensional input pattern vector, $m = 1, 2, ..., M$, and K is the number of hidden units. M is the number of classes (outputs). w_{mj} is the weight connecting the jth hidden unit to the mth output node. b_m is the bias. w_{m0} is the weight connecting the bias and the mth output node.

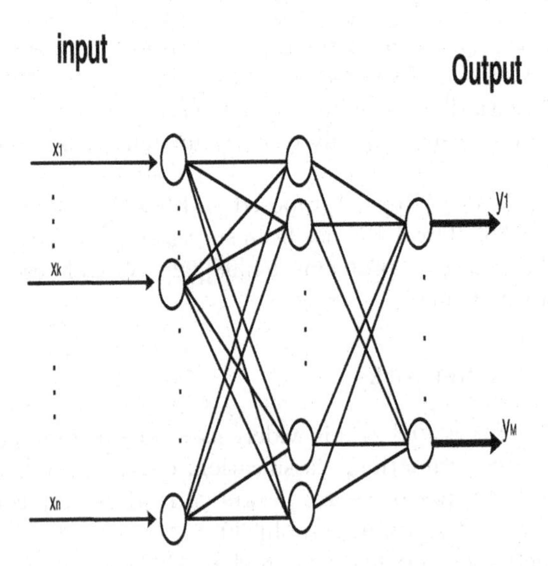

Fig. 1.3. Architecture of an RBF neural network. (© 2005 IEEE) We thank the IEEE for allowing the reproduction of this figure, first appeared in [104].

The radial basis activation function $\o(x)$ of the RBF neural network distinguishes it from other types of neural networks. Several forms of activation functions have been used in applications:

1.
$$\o(x) = \mathrm{e}^{-x^2/2\sigma^2}, \tag{1.6}$$

2.
$$\o(x) = (x^2 + \sigma^2)^{-\beta},\ \beta > 0, \tag{1.7}$$

3.
$$\o(x) = (x^2 + \sigma^2)^{\beta},\ \beta > 0, \tag{1.8}$$

4.
$$\o(x) = x^2 ln(x); \tag{1.9}$$

here σ is a parameter that determines the smoothness properties of the interpolating function.

The Gaussian kernel function and the function (Eq. (1.7)) are localized functions with the property that $\o \rightarrow 0$ as $|x| \rightarrow \infty$. One-dimensional Gaussian function is shown in Fig. 1.4. The other two functions (Eq. (1.8), Eq. (1.9)) have the property that $\o \rightarrow \infty$ as $|x| \rightarrow \infty$.

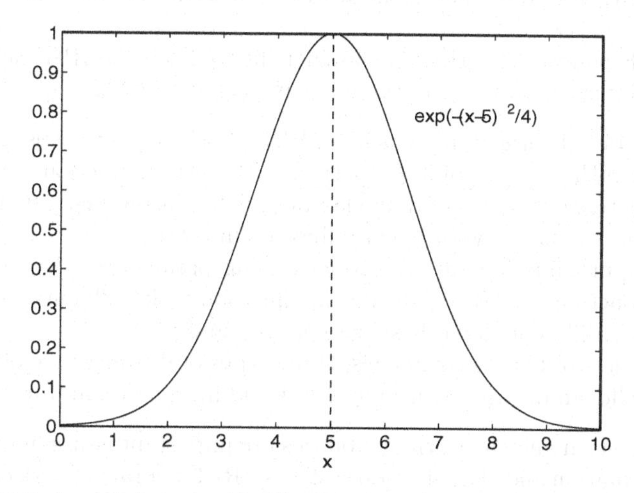

Fig. 1.4. Bell-shaped Gaussian Profile: The kernel possesses the highest response at the center $x = 5$ and degrades to zero quickly

In this book, the activation function of RBF neural networks is the Gaussian kernel function. $\o_j(X)$ is the activation function of the jth hidden unit:

$$\o_j(X) = \mathrm{e}^{-||\mathbf{X}-\mathbf{C_j}||^2/2\sigma_j{}^2}, \tag{1.10}$$

where $\mathbf{C}_j$ and σ_j are the center and the width for the jth hidden unit, respectively, which are adjusted during learning. When calculating the distance between input patterns and centers of hidden units, Euclidean distance measure is employed in most RBF neural networks.

RBF neural networks are able to make an exact interpolation by passing through every data point $\{X_i, Y_i\}$. In practice, noise is often present in data sets and an exact interpolation may not be desirable. Proomhead and Lowe [33] proposed a new RBF neural network model to reduce computational complexity, i.e., the number of radial basis functions. In [219], a smooth interpolating function is generated by the RBF network with a reduced number of radial basis functions.

Consider the following two major function approximation problems:

(a) target functions are known. The task is to approximate the known function by simpler functions, such as Gaussian functions,

(b) target functions are unknown but a set of samples $\{x, y(x)\}$ are given. The task is to approximate the function y.

RBF neural networks with free adjustable radial basis functions or prototype vectors are universal approximators, which can approximate any continuous function with arbitrary precision if there are sufficient hidden neurons [237][282]. The domain of y can be a finite set or an infinite set. If the domain of y is a finite set, RBF neural networks deal with classification problems [241].

The RBF neural network as a classifier differs from the RBF neural network as an interpolation tool in the following aspects [282]:

1. The number of kernel functions in an RBF classifier model is usually much fewer than the number of input patterns. The kernel functions are located in the centers of clusters of RBF classifiers. The clusters separate the input space into subspaces with hyper-ellipse boundaries.
2. In the approximation task, a global scaling parameter σ is used for all kernel functions. However, in the classification task, different σ's are employed for different radial basis kernel functions.
3. In RBF network classifier models, three types of distances are often used. The Euclidean distance is usually employed in function approximation.

Generalization and the learning abilities are important issues in both function approximation and classification tasks. An RBF neural network can attain no errors for a given training data set if the RBF network has as many hidden neurons as the training patterns. However, the size of the network may be too large when tackling large data sets and the generalization ability of such a large RBF network may be poor. Smaller RBF networks may have better generalization ability; however, too small a RBF neural network will perform poorly on both training and test data sets. It is desirable to determine a training method which takes the learning ability and the generalization ability into consideration at the same time.

Three training schemes for RBF networks [282] are as follows:

- One-stage training
  In this training procedure, only the weights connecting the hidden layer and the output layer are adjusted through some kind of supervised methods, e.g., minimizing the squared difference between the RBF neural network's output and the target output. The centers of hidden neurons are subsampled from the set of input vectors (or all data points are used as centers) and, typically, all scaling parameters of hidden neurons are fixed at a predefined real value [282] typically.
- Two-stage training
  Two-stage training [17][22][36][264] is often used for constructing RBF neural networks. At the first stage, the hidden layer is constructed by selecting the center and the width for each hidden neuron using various clustering algorithms. At the second stage, the weights between hidden neurons and output neurons are determined, for example by using the linear least square (LLS) method [22]. For example, in [177][280], Kohonen's learning vector quantization (LVQ) was used to determine the centers of hidden units. In [219][281], the k-means clustering algorithm with the selected data points as seeds was used to incrementally generate centers for RBF neural networks. Kubat [183] used C.4.5 to determine the centers of RBF neural networks. The width of a kernel function can be chosen as the standard deviation of the samples in a cluster. Murata *et al.* [221] started with a sufficient number of hidden units and then merged them to reduce the size of an RBF neural network. Chen *et al.* [48][49] proposed a constructive method in which new RBF kernel functions were added gradually using an orthogonal least square learning algorithm (OLS). The weight matrix is solved subsequently [48][49].
- Three-stage training
  In a three-stage training procedure [282], RBF neural networks are adjusted through a further optimization after being trained using a two-stage learning scheme. In [73], the conventional learning method was used to generate the initial RBF architecture, and then the conjugate gradient method was used to tune the architecture based on the quadratic loss function.

An RBF neural network with more than one hidden layer is also presented in the literature. It is called the multi-layer RBF neural network [45]. However, an RBF neural network with multiple layers offers little improvement over the RBF neural network with one hidden layer. The inputs pass through an RBF neural network and form subspaces of a local nature. Putting a second hidden layer after the first hidden layer will lead to the increase of the localization and the decrease of the valid input signal paths accordingly [138]. Hirasawa *et al.* [138] showed that it was better to use the one-hidden-layer RBF neural network than using the multi-layer RBF neural network.

Given N patterns as a training data set, the RBF neural network classifier may obtain 100% accuracy by forming a network with N hidden units, each of which corresponds to a training pattern. However, the 100% accuracy in the training set usually cannot lead to a high classification accuracy in the test data set (the unknown data set). This is called the generalization problem. An important question is: 'how do we generate an RBF neural network classifier for a data set with the fewest possible number of hidden units and with the highest possible generalization ability?'.

The number of radial basis kernel functions (hidden units), the centers of the kernel functions, the widths of the kernel functions, and the weights connecting the hidden layer and the output layer constitute the key parameters of an RBF classifier. The question mentioned above is equivalent to how to optimally determine the key parameters. Prior knowledge is required for determining the so-called 'sufficient number of hidden units'. Though the number of the training patterns is known in advance, it is not the only element which affects the number of hidden units. The data distribution is another element affecting the architecture of an RBF neural network. We explore how to construct a compact RBF neural network in the latter part of this book.

1.2.4 Support Vector Machines

Support vector machines (SVMs) [62][326][327] have been widely applied to pattern classification problems [46][79][148][184][294] and non-linear regressions [230][325]. SVMs are usually employed in pattern classification problems. After SVM classifiers are trained, they can be used to predict future trends. We note that the meaning of the term *prediction* is different from that in some other disciplines, e.g., in time-series prediction where prediction means guessing future trends from past information. Here, 'prediction' means supervised classification that involves two steps. In the first step, an SVM is trained as a classifier with a part of the data in a specific data set. In the second step (i.e., prediction), we use the classifier trained in the first step to classify the rest of the data in the data set.

The SVM is a statistical learning algorithm pioneered by Vapnik [326][327]. The basic idea of the SVM algorithm [29][62] is to find an optimal hyper-plane that can maximize the margin (a precise definition of margin will be given later) between two groups of samples. The vectors that are nearest to the optimal hyper-plane are called support vectors (vectors with a circle in Fig. 1.5) and this algorithm is called a support vector machine. Compared with other algorithms, SVMs have shown outstanding capabilities in dealing with classification problems. This section briefly describes the SVM.

Linearly Separable Patterns

Given l input vectors $\{\mathbf{x}_i \in R^n, i = 1, ..., l\}$ that belong to two classes, with desired output $y_i \in \{-1, 1\}$, if there exists a hyper-plane

$$\mathbf{w}^{\mathrm{T}}\mathbf{x} + b = 0 \tag{1.11}$$

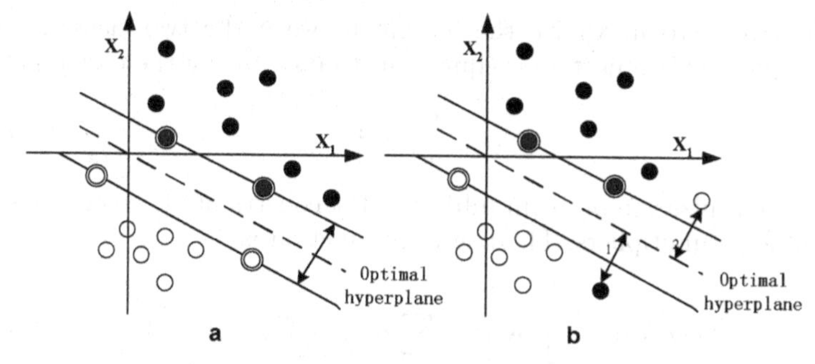

Fig. 1.5. An optimal hyper-plane for classification in a two-dimensional case, for (a) linearly separable patterns and (b) linearly non-separable patterns.

that separates the two classes, that is,

$$\mathbf{w}^{\mathrm{T}}\mathbf{x}_i + b \geq 0, \text{ for all } i \text{ with } y_i = +1, \tag{1.12}$$

$$\mathbf{w}^{\mathrm{T}}\mathbf{x}_i + b < 0, \text{ for all } i \text{ with } y_i = -1, \tag{1.13}$$

then we say that these patterns are *linearly separable.* Here $\mathbf{w}$ is a weight vector and b is a bias. By rescaling $\mathbf{w}$ and b properly, we can change the two inequalities above to:

$$\mathbf{w}^{\mathrm{T}}\mathbf{x}_i + b \geq 1, \text{ for all } i \text{ with } y_i = +1, \tag{1.14}$$

$$\mathbf{w}^{\mathrm{T}}\mathbf{x}_i + b \leq -1, \text{ for all } i \text{ with } y_i = -1. \tag{1.15}$$

Or,

$$y_i(\mathbf{w}^{\mathrm{T}}\mathbf{x}_i + b) \geq -1. \tag{1.16}$$

There are two parallel hyper-planes:

$$H1\text{: } \mathbf{w}^{\mathrm{T}}\mathbf{x} + b = 1, \tag{1.17}$$

$$H2\text{: } \mathbf{w}^{\mathrm{T}}\mathbf{x} + b = -1. \tag{1.18}$$

The distance ρ between $H1$ and $H2$ is defined as the *margin* between the two classes (Fig. 1.5a). According to the standard result of the distance between the origin and a hyper-plane, we can figure out that the distances between the origin and $H1$ and $H2$ are $|b-1|/||\mathbf{w}||$ and $|b+1|/||\mathbf{w}||$, respectively. The sum of these two distances is ρ, because $H1$ and $H2$ are parallel. Therefore,

$$\rho = 2/||\mathbf{w}||. \tag{1.19}$$

The objective is to maximize the margin between the two classes, i.e., to minimize $||\mathbf{w}||$. This objective is equivalent to minimizing the cost function:

$$\psi = \frac{1}{2}||\mathbf{w}||^2. \tag{1.20}$$

Then, this optimization problem subject to the constraint (1.16) can be solved using Lagrange multipliers. The Lagrange function is

$$L(\mathbf{w}, b, \alpha) = \frac{1}{2}||\mathbf{w}||^2 - \sum_{i=1}^{l} \alpha_i [y_i(\mathbf{w}^{\mathrm{T}}\mathbf{x}_i + b) - 1], \tag{1.21}$$

where $\alpha_i, i = 1, 2, ..., l$ are the Lagrange multipliers. Differentiating this Lagrange function, we obtain

$$\frac{\partial L(\mathbf{w}, b, \alpha)}{\partial \mathbf{w}} = \mathbf{0}, \tag{1.22}$$

$$\frac{\partial L(\mathbf{w}, b, \alpha)}{\partial \alpha} = 0. \tag{1.23}$$

Considering the Wolfe's dual [89], we can obtain a dual problem of the primal one:

$$\text{maximize: } Q(\alpha) = \sum_{i=1}^{l} \alpha_i - \frac{1}{2} \sum_{i=1}^{l} \sum_{j=1}^{l} \alpha_i \alpha_j y_i y_j \mathbf{x}^{\mathrm{T}}\mathbf{x}, \tag{1.24}$$

subject to:

$$\sum_{i=1}^{l} \alpha_i y_i = 0, \tag{1.25}$$

$$\alpha_i \geq 0. \tag{1.26}$$

From this dual problem, the optimal weight vector, i.e., $\mathbf{w}_o$ and the optimal Lagrange multipliers, i.e., $\alpha_{o,i}$ of the optimal hyper-plane can be obtained:

$$\mathbf{w}_o = \sum_{i=1}^{l} \alpha_{o,i} y_i \mathbf{x}_i. \tag{1.27}$$

Linearly Non-separable Patterns

If the vectors $\{\mathbf{x}_i \in R^n, i = 1, ..., l\}$ cannot be linearly separated, we would like to slacken the constraints described by (1.16). Here we introduce a group of slack variables, i.e., ξ_i:

$$y_i(\mathbf{w}^{\mathrm{T}}\mathbf{x}_i + b) \geq 1 - \xi_i, \tag{1.28}$$

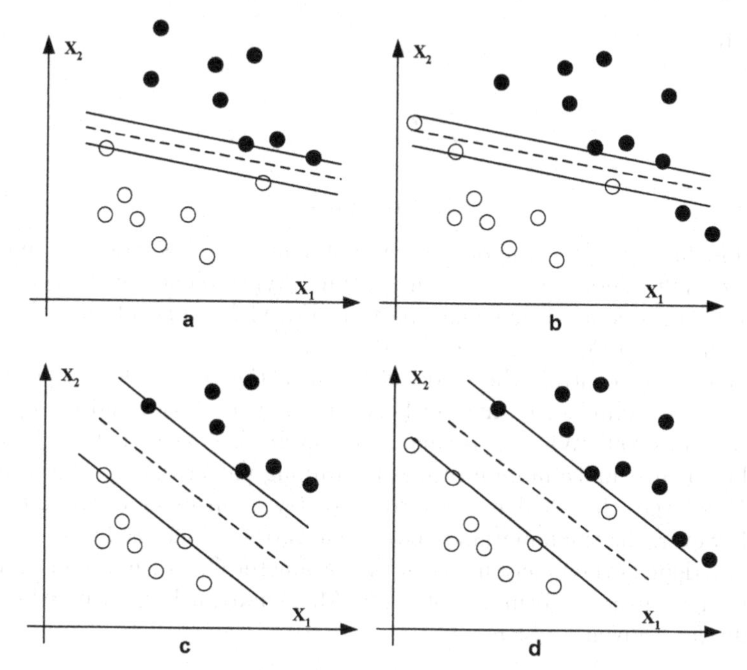

Fig. 1.6. The influence of C on the performance of the classifier. (a) a classifier with a large C (small margin); (b) an overfitting classifier; (c) a classifier with a small C (large margin); (d) a classifier with a proper C.

$$\xi_i \geq 0. \tag{1.29}$$

In fact, ξ_i is the distance between the training example $\mathbf{x}_i$ and the optimal hyper-plane (Fig. 1.5b). For $0 \leq \xi_i \leq 1$, $\mathbf{x}_i$ falls in the region between the two hyper-planes, i.e., $H1$ and $H2$, but on the correct side of the optimal hyper-plane. However, for $\xi_i > 1$, $\mathbf{x}_i$ falls on the wrong side of the optimal hyper-plane.

Since it is expected that the optimal hyper-plane can maximize the margin between the two classes and minimize the errors, the cost function from Eq. (1.20) is rewritten:

$$\psi = \frac{1}{2}||\mathbf{w}||^2 + C\sum_{i=1}^{l} \xi_i, \tag{1.30}$$

where C is a positive factor. This cost function must satisfy the constraints Eq. (1.28) and Eq. (1.29). There is also a dual problem:

$$\text{maximize: } Q(\alpha) = \sum_{i=1}^{l} \alpha_i - \frac{1}{2}\sum_{i=1}^{l}\sum_{j=1}^{l} \alpha_i \alpha_j y_i y_j \mathbf{x}^{\mathrm{T}}\mathbf{x}, \tag{1.31}$$

subject to:

$$\sum_{i=1}^{l} \alpha_i y_i = 0, \tag{1.32}$$

$$C \geq \alpha_i \geq 0. \tag{1.33}$$

From this dual problem, the optimal weight vector, i.e., $\mathbf{w}_o$ and the optimal Lagrange multipliers, i.e., $\alpha_{o,i}$ of the optimal hyper-plane can be obtained. They are the same as their counterparts in Eq. (1.27), except that the constraints change to Eq. (1.32) and (1.33).

In general, C controls the trade-off between the two goals of the binary SVM, i.e., to maximize the margin between the two classes and to separate the two classes well. When C is small, the margin between the two classes is large, but it may make more mistakes in training patterns. Or, alternatively, when C is large, the SVM is likely to make fewer mistakes in training patterns; however, the small margin makes the network vulnerable for overfitting. Figure 1.6 depicts the functionality of the parameter C, which has a relatively large impact on the performance of the SVM. Usually, it is determined experimentally for a given problem.

A Binary Non-linear SVM Classifier

According to [65], if a non-linear transformation can map the input feature space into a new feature space whose dimension is high enough, the classification problem is more likely to be linearly solved in this new high-dimensional space. In view of this theorem, the *non-linear* SVM algorithm performs such a transformation to map the input feature space to a new space with much higher dimension. Actually, other kernel learning algorithms, such as radial basis function (RBF) neural networks, also perform such a transformation for the same reason. After the transformation, the features in the new space are classified using the optimal hyper-plane we constructed in the previous sections. Therefore, using this non-linear SVM to perform classification includes the following two steps:

1. Mapping the input space into a much higher dimensional space with a non-linear kernel function.
2. Performing classification in the new high-dimensional space by constructing an optimal hyper-plane that is able to maximize the margin between the two classes.

Combining the transformation and the linear optimal hyper-plane, we formulate the mathematical descriptions of this non-linear SVM as follows.

It is supposed to find the optimal values of weight vector $\mathbf{w}$ and bias b such that they satisfy the constraint:

$$y_i(\mathbf{w}^{\mathrm{T}}\phi(\mathbf{x}_i) + b) \geq 1 - \xi_i, \quad (1.34)$$

$$\xi_i \geq 0. \quad (1.35)$$

where $\phi(\mathbf{x}_i)$ is the function mapping the ith pattern vector to a potentially much higher dimensional feature space. The weight vector $\mathbf{w}$ and the slack variables ξ_i should minimize the cost function:

$$\psi = \frac{1}{2}||\mathbf{w}||^2 + C\sum_{i=1}^{l}\xi_i, \quad (1.36)$$

This optimization problem is very similar to the problem we have dealt with using a linear optimal hyper-plane. The only difference is that the input vectors $\mathbf{x}_i$ have been replaced by $\phi(\mathbf{x}_i)$.

To solve this optimization problem, a similar procedure is followed as before. Through constructing the Lagrange function and differentiating it, a dual problem is obtained as below:

$$\text{maximize: } Q(\alpha) = \sum_{i=1}^{l}\alpha_i - \frac{1}{2}\sum_{i=1}^{l}\sum_{j=1}^{l}\alpha_i\alpha_j y_i y_j K(\mathbf{x}_i, \mathbf{x}_j), \quad (1.37)$$

subject to:

$$\sum_{i=1}^{l}\alpha_i y_i = 0, \quad (1.38)$$

$$C \geq \alpha_i \geq 0, \quad (1.39)$$

where $K(\mathbf{x}_i, \mathbf{x}_j)$ is the kernel function:

$$K(\mathbf{x}_i, \mathbf{x}_j) = \phi(\mathbf{x}_i)^{\mathrm{T}}\phi(\mathbf{x}_j). \quad (1.40)$$

From this dual problem, the optimal weight vector i.e., $\mathbf{w}_o$ and the optimal Lagrange multipliers, i.e., $\alpha_{o,j}$ of the optimal hyper-plane can be obtained:

$$\mathbf{w}_o = \sum_{i=1}^{l}\alpha_{o,i} y_i \mathbf{x}_i. \quad (1.41)$$

The optimal hyper-plane that discriminates different classes is:

$$\mathbf{w}_o^{\mathrm{T}}\phi(\mathbf{x}) + b = 0. \quad (1.42)$$

One of the most commonly used kernel functions is the polynomial kernel:

$$K(\mathbf{x}, \mathbf{x}_i) = (\mathbf{x}^{\mathrm{T}}\mathbf{x}_i + 1)^p, \quad (1.43)$$

where p is a constant specified by users. Another kind of widely used kernel function is the radial basis function:

$$K(\mathbf{x}, \mathbf{x}_i) = e^{-\gamma||\mathbf{x}-\mathbf{x}_i||^2}, \tag{1.44}$$

where γ is also a constant specified by users. According to its mathematical description, the structure of an SVM is shown in Fig. 1.7.

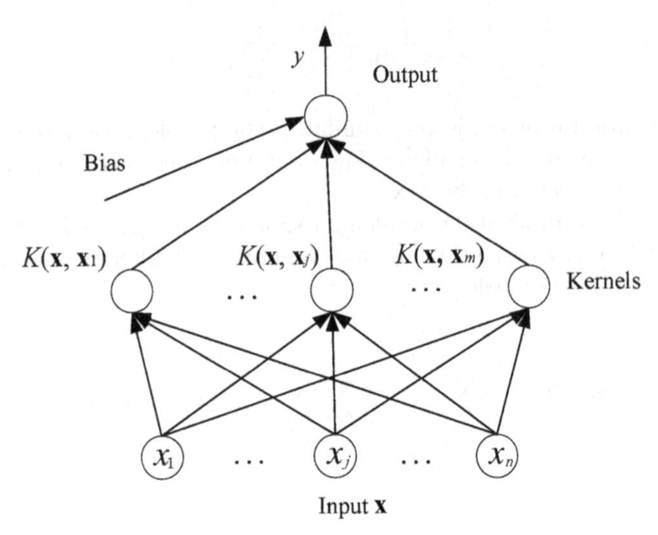

Fig. 1.7. The structure of an SVM.

1.2.5 Genetic Algorithms

Genetic algorithms (GAs) are motivated by the natural evolutionary process. The basic concepts in GAs are as follows. Solutions of the problem at hand are encoded in chromosomes or individuals. An initial population of individuals is generated at random or heuristically. The operators in GAs include selection, crossover, and mutation. To generate a new generation, chromosomes are selected according to their fitness scores, i.e., a predefined quality criterion used for evaluating solutions of a problem. The selection operator gives preference to better individuals as parents for the next generation. The crossover operator and the mutation operator are used to generate offspring from the parents. A crossover site is randomly chosen in the parents. The two bit strings in the two individuals are exchanged up to the crossover site. For example, suppose that parents $I_1 = 0001100$ and $I_2 = 1110000$ are selected for generating new offspring. After applying the crossover operator, we obtain $I_1' = 0010100$ and $I_2' = 1101000$ with two crossover points at the third and fourth bits. By exchanging portions of good individuals, crossover may produce even better individuals. The mutation operator is used to prevent premature convergence to local optima. It is implemented by flipping bits at random with a mutation probability.

GAs are specially useful under the following circumstances:

- the problem space is large, complex;
- prior knowledge is scarce;
- it is difficult to determine a machine learning model to solve the problem due to complexities in constraints and objectives;
- traditional search methods perform badly.

The steps to apply the basic GA as a problem-solving model are as follows:

1. figure out a way to encode solutions of the problem according to domain knowledge and required solution quality;
2. randomly generate an initial population of chromosomes which corresponds to solutions of the problem;
3. calculate the fitness of each chromosome in the population pool;
4. select two parental chromosomes from the population pool to produce offspring by crossover and mutation operators;
5. go to step 3, and iterate until an optimal solution is found.

The basic genetic algorithm is simple but powerful in solving problems in various areas. In addition, the basic GA could be modified to meet requirements of diverse problems by tuning the basic operators. For a detailed discussion of variations of the basic GA, as well as other techniques in a broader category called evolutionary computation, see text books, such as [10][86].

1.3 How This Book is Organized

In Chap. 1, data mining tasks and conventional data mining methods are introduced. Classification and clustering tasks are explained, with emphasis on the classification task. An introduction to data mining methods is presented.

In Chap. 2, a wavelet multi-layer perceptron neural network is described for predicting temporal sequences. The multi-layer perceptron neural network has its input signal decomposed to various resolutions using a wavelet transformation. The time frequency information which is normally hidden is exposed by the wavelet transformation. Based on the wavelet transformation, less important wavelets are eliminated. Compared with the conventional MLP network, the wavelet MLP neural network has less performance swing sensitivity to weight initialization. In addition, we describe a cost-sensitive MLP in which errors in prediction are biased towards 'important' classes. Since different prediction errors in different classes usually lead to different costs, it is worthwhile discussing the cost-sensitive problem. In experimental results, it is shown that the recognition rates for the 'important' classes (with higher cost) are higher than the recognition rates for the 'less important' classes.

In Chap. 3, the FNN is described. This FNN that we proposed earlier combines the powerful features of initial fuzzy model self-generation, fast input selection, partition validation, parameter optimization, and rule-base simplification. The structure and learning procedure are introduced first. Then, we describe the implementation and functionality of the FNN. Synthetic databases and microarray data are used to demonstrate the fuzzy neural network proposed earlier [59][349]. Experimental results are compared with the pruned feedforward crisp neural network and decision tree approaches.

Chapter 4 describes how to construct an RBF neural network that allows for large overlaps between clusters with the same class label, which reduces the number of hidden units without degrading the accuracy of the RBF neural network. In addition, we describe a new method dealing with unbalanced data. The method is based on the modified RBF neural network. Weights inversely proportional to the number of patterns of classes are given to each class in the mean squared error (MSE) function.

In Chap. 5, DDR methods, including feature selection and feature extraction techniques, are reviewed first. A novel algorithm for attribute importance ranking, i.e., the separability and correlation measure (SCM), is then presented. Class-separability measure and attribute-correlation measure are weighted to produce a combined evaluation for relative attribute importance. The top-down search and the bottom-up search are explored, and their difference in attribute ranking is presented. The attribute ranking algorithm with class information is compared with other attribute ranking methods. Data dimensionality is reduced based on attribute ranking results.

Data dimensionality reduction is then performed by combining the SCM method and RBF classifiers. In the DDR method, there are a fewer number of candidate feature subsets to be inspected compared with other methods, since attribute importance is ranked first by the SCM method. The size of a data set is reduced and the architecture of the RBF classifier is simplified. Experimental results show the advantages of the DDR method.

In Chap. 6, reviews of existing class-dependent feature selection techniques are presented first. The fact that different features might have different discrimination capabilities for separating one class from the other classes is adopted. For a multi-class classification problem, each class has its own specific feature subset as inputs of the RBF neural network classifier. The novel class-dependent feature selection algorithm is based on RBF neural networks and the genetic algorithm (GA).

In Chap. 7, reviews of rule extraction work in the literature are presented first. Several new rule extraction methods are described based on the simplified RBF neural network classifier in which large overlaps between clusters of the same class are allowed. In the first algorithm, A GA combined with an RBF neural network is used to extract rules. The GA is used to determine the intervals of each attribute as the premise of the rules. In the second algorithm, rules are extracted directly based on simplified RBF neural networks using gradient descent. In the third algorithm, the DDR technique is combined with rule extraction. Rules with a fewer number of premises (attributes) and higher rule accuracy are obtained. In the fourth algorithm, class-dependent feature selection is used as a preprocessing procedure of rule extraction. The results from the four algorithms are compared with other algorithms.

In Chap. 8, a hybrid neural network predictor is described for protein secondary structure prediction (PSSP). The hybrid network is composed of the RBF neural network and the MLP neural network. Experiments show that the performance of the hybrid network has reached a comparable performance with the existing leading method.

In Chap. 9, support vector machine classifiers are used to deal with two bioinformatics problems, i.e., cancer diagnosis based on gene expression data and protein secondary structure prediction.

Chapter 10 describes a rule extraction algorithm RulExSVM that we proposed earlier [108]. Decisions made by a non-linear SVM classifier are decoded into linguistic rules based on the support vectors and decision functions according to a geometrical relationship.

2

MLP Neural Networks for Time-Series Prediction and Classification

2.1 Wavelet MLP Neural Networks for Time-series Prediction

In this chapter, we investigate the effectiveness of wavelet multi-layer perceptron (MLP) neural networks (NNs) for temporal sequence prediction. It is essentially a neural network with the input signal decomposed to various resolutions using wavelet transforms. Wavelet transforms can expose time-frequency information that is normally hidden. We show that the wavelet MLP network provides a prediction performance comparable to the conventional MLP. After the less important inputs are eliminated, the wavelet MLP shows more consistent performance for different weight initializations in comparison to the conventional MLP [303][332].

2.1.1 Introduction to Wavelet Multi-layer Neural Network

A time-series is a sequence of data that vary with time, for example, the daily average temperature from the year 1995 to 2005. The task of time-series prediction is to forecast future trend using the past values in the time-series.

There exist many approaches to time-series prediction. The oldest and most studied method, a linear autoregression (AR), is to fit the data using the following equation [47]:

$$y(k) = \sum_{i=1}^{T} a(i)y(k-i) + e(k) = \hat{y}(k) + e(k), \tag{2.1}$$

where $y(k)$ is the actual value of the time-series at time step k, $a(i)$ is the weight for time step i, and $e(k)$ is the prediction error. $\hat{y}(k)$ is the predicted value of $y(k)$.

AR represents $y(k)$ as a weighted sum of past values of the sequence. This model can provide good performance only when the system under investigation is linear or nearly linear. However, the performance may be very poor for cases in which the system dynamics is highly non-linear.

NNs have demonstrated great potential for time-series prediction where the system dynamics is non-linear. Lapedes and Farber [186] first studied non-linear signal prediction using an MLP. It led to an explosive increase in research activities in examining the approximation capabilities of MLPs [132][340].

Artificial NNs were developed to emulate the human brain that is powerful, flexible, and efficient. However, conventional networks process the signal only at its finest resolution, which is not the case for the human brain. For example, the retinal image is likely to be processed in separate frequency channels [205].

The introduction of wavelet decomposition [204][293] provides a new tool for approximation. Inspired by both the MLP and wavelet decomposition, Zhang and Benveniste [357] invented a new type of network, call a *wavelet network*. This has caused rapid development of a new breed of neural network models integrated with wavelets. Most researchers used wavelets as radial basis functions that allow hierarchical, multi-resolution learning of input-output maps from experimental data [15][52]. Liang and Page [193] proposed a new learning concept and paradigm for a neural network, called multi-resolution learning based on multi-resolution analysis in wavelet theory.

In this chapter, we use wavelets to break the signal down into its multiresolution components before feeding them into an MLP. We show that the wavelet MLP neural network is capable of utilizing the time-frequency information to improve its consistency in performance.

2.1.2 Wavelet

The wavelet theory provides a unified framework for a number of techniques that had been developed independently for various signal-processing applications, e.g., multi-resolution signal processing used in computer vision, subband coding developed for speech and image compression, and wavelet series expansions developed in applied mathematics. In this section, we will concentrate on the multi-resolution approximation to be discussed in this chapter.

Multi-resolution

Wavelet ψ can be constructed such that the dilated and translated family

$$\{\psi_{j,i}(t) = \sqrt{2^j}\psi(2^j(t-i))\}_{(j,i)} \in Z^1, \tag{2.2}$$

where ψ (mother wavelet) is an orthonormal basis of $L^2(\mathbf{R})$ and $L^2(\mathbf{R})$ denotes the vector space of square-integrable, one-dimensional functions $f(x)$. Let $\mathbf{V}_j$ denote a closed subspace in $L^2(\mathbf{R})$. Orthogonal wavelets dilated by 2^j carry signal variations at resolution 2^j. Thus a wavelet can be used to compute the approximation of the signal at various resolutions with orthogonal projections on different spaces $\{\mathbf{V}_j\}_{j \in Z}$. Each subspace contains the approximation of all functions $f(x)$ at resolution 2^j. The approximation of the signal at resolution 2^{j+1} contains all information necessary to compute the signal at the lower resolution. Thus, they are a set of nested vector subspaces,

$$\cdots \subset V_j \subset V_{j+1} \subset V_{j+2} \subset \cdots \quad (2.3)$$

When computing the approximation of function f at resolution 2^j, some information about f is lost. As the resolution increases to infinity, the approximate signal converges to the original signal. When the resolution approaches zero, the signal vanishes. If P_{vj} denotes the orthogonal projection operator from $L^2(\mathbf{R})$ onto $\mathbf{V}_j$,

$$\lim_{j \to -\infty} \|P_{vj}, f\| = 0. \quad (2.4)$$

On the other hand, when the resolution approaches $+\infty$, the signal approximation converges to the original signal:

$$\lim_{j \to +\infty} \|f - P_{vj}, f\| = 0. \quad (2.5)$$

The limit (2.5) guarantees that the original signal can be reconstructed using decomposed signals at a lower resolution.

Signal Decomposition

A tree algorithm can be used for computing the wavelet transform by using the wavelet coefficients as filter coefficients. Assume that vector s^m represents the sampled signal f at the finest resolution 2^m. A low-pass filter L is employed to produce a coarser approximation at resolution 2^{m-1}. Thus,

$$s^{j-1} = Ls^j \quad j = 1, 2, ..., m. \quad (2.6)$$

The detailed signal d^j at resolution 2^j is obtained by applying a high-pass filter H to s^j. That is,

$$d^{j-1} = Hs^j, \quad j = 1, 2, ..., m. \quad (2.7)$$

The process can be repeated to produce signals at any desired resolution (Fig. 2.1).

The signal can be reconstructed using two synthesis filters L^* and H^* (the transposed matrices of L and H, respectively). The reconstruction is given by Fig. 2.2.

Hence, any original signal can be represented as

$$f = s^m = s^0 + d^0 + d^1 + \cdots + d^{m-1} + d^m. \quad (2.8)$$

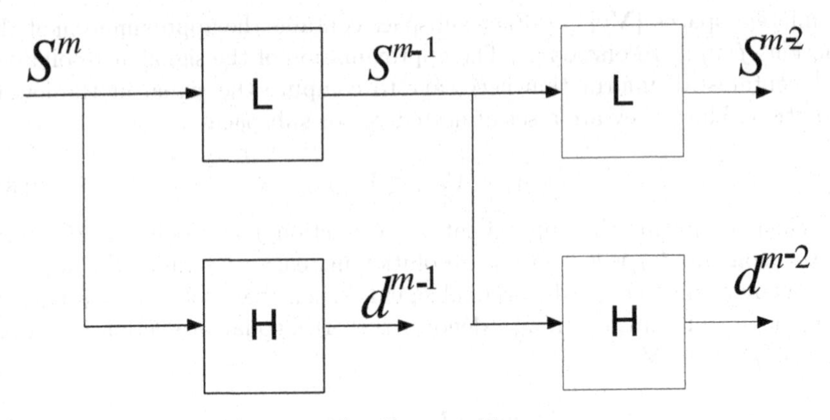

Fig. 2.1. The decomposition process.

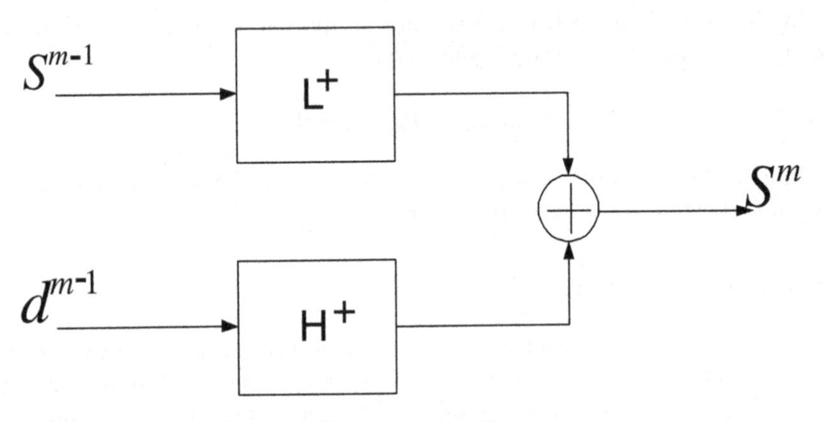

Fig. 2.2. The reconstruction process.

2.1.3 Wavelet MLP Neural Network

Figure 2.3 shows the wavelet MLP neural network used in this chapter. The input signal is passed through a tapped delay line to create short-term memory that retains aspects of the input sequence relevant to making predictions. This is similar to a time-lagged MLP except that the delayed data is not sent directly into the network. Instead, it is decomposed by a wavelet transform to form the input of the MLP. Figure 2.4 shows an example of two-level decomposition of the tapped delay data x. Data x is decomposed to coarser (CA1) and detailed (CD1) approximations. The coarser approximation (CA1) is further decomposed into its coarser (CA2) and detailed (CD2) approximations.

Furthermore, we are looking into the possibility of discarding certain wavelet-decomposed data that is of little use in the mapping of input to output. The mapping is expected to be highly non-linear and dependent on the characteristics of the individual signal.

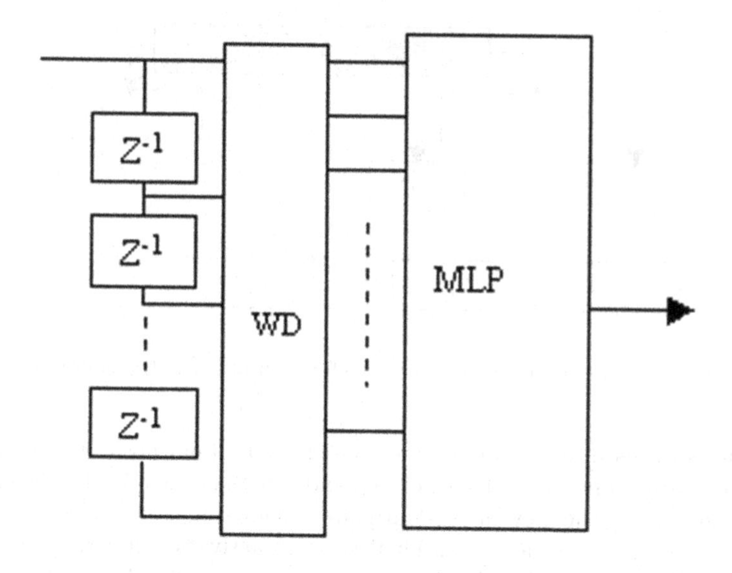

Fig. 2.3. Model of Network (WD=Wavelet Decomposition).

Let

$$s_i = \sum_{j=1}^{n} |w_{ij}|, \tag{2.9}$$

represent the importance of input x_i, where w_{ij} is the weighting of input i to neuron j and n is the number of hidden neurons.

$$s_i' = \frac{s_i'}{\max(s_i)} \tag{2.10}$$

serves as an indicator of the relative importance of input x_i. Here s_i' is the normalized input strength, $\max(s_i)$ is the maximum of s_1, s_2, ..., s_I, and I is the number of inputs.

Input points having small s_i' will be considered to be unimportant and may be discarded without affecting the prediction performance.

2.1.4 Experimental Results

The Mackey-Glass time-series is frequently used as a benchmark in time-series prediction.

The Mackey-Glass time-delay differential equation is defined by

$$\frac{dx(t)}{dt} = \frac{0.2x(t-\phi)}{(1+x(t-\phi))^{10}} - 0.1x(t). \tag{2.11}$$

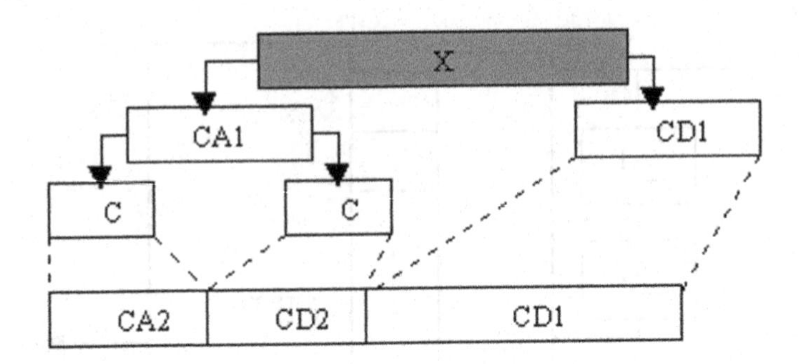

Fig. 2.4. The two-level decomposition to form an input to the neural network.

The MLP used in our simulations consists of an input layer, a hidden layer of two neurons, and one output neuron, and is trained by a back-propagation algorithm using the Levenberg-Marquardt algorithm for fast optimization [127]. All neurons use a conventional sigmoid activation function; however, the output neuron employed a linear activation function as frequently used in forecasting applications.

In order to compare our result, the normalized mean squared error (NMSE) is used to assess forecasting performance. The NMSE is computed as

$$\text{NMSE} = \frac{1}{\sigma^2} \frac{1}{N} \sum_{t=1}^{N} [x(t) - \hat{x}(t)]^2, \tag{2.12}$$

where $x(t)$ is the actual value of the time-series, $\hat{x}(t)$ is the predicted value of $x(t)$, σ^2 is the variance of the time-series over the predicting duration, and N is the number of elements.

The data is divided into three parts for the training, validation, and testing, respectively. The training data is of length 220, followed by validation and testing data, each of length 30. Validation NMSE is evaluated every 20 epochs. When there is an increase in the validation NMSE, training stops. Test data is used to test the generalization performance of the network and is not used by the network during training or validation.

Early stopping by monitoring validation error often shows multiple minima as a function of training time and results are also sensitive to the weight initialization [340]. In order to have a fair comparison, simulations are carried out for each network with different random weight initializations over 100 trials. The 50 lowest NMSEs are kept for calculations of mean and standard deviation, which are then used for comparisons.

The simulations indicate that the input points 1, 4, and 5 are consistently less important than other inputs (Fig. 2.5). Simulations are re-run after these less important inputs are eliminated. This results in a network of size 17:2:1 (seventeen inputs, two hidden neurons and one output neuron). We denote

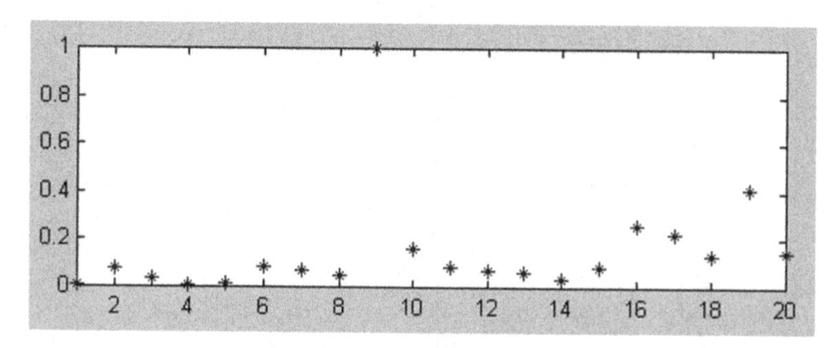

Fig. 2.5. Distribution of relative importance of 20 inputs for the wavelet MLP network with decomposition level one in one of the simulations, which is similar to the results in other simulations.

this wavelet MLP neural network by (17:2:1) = (20:2:1)-[1,4,5]. Simulations are carried out on other network sizes to reduce the number of inputs when possible.

Table 2.1 shows that a wavelet MLP network provides a prediction performance comparable to the conventional MLP. After less important inputs are eliminated, the wavelet MLP shows a more consistent performance for different weight initializations than the conventional MLP does.

Table 2.1. Result of the three networks with different architectures.

		Normalized mean square error (NMSE)		
Architecture	Type	Mean	Standard deviation	Minimum
	Conventional MLP	0.063	0.016	0.0047
20:2:1	Wavelet MLP	0.054	0.0026	0.036
	Conventional MLP	0.043	0.0093	0.0069
17:2:1	Wavelet MLP	0.045	1.38×10^{-6}	0.045
	(20:2:1)-[1,4,5]	0.027	8.051×10^{-6}	0.027
	Conventional MLP	0.021	0.0107	0.0017
12:2:1	Wavelet MLP	0.032	0.0030	0.013
	Conventional MLP	0.015	0.0078	0.0017
	Wavelet MLP	0.032	0.0029	0.017
11:2:1	(12:2:1)-[7]	0.021	0.00064	0.017

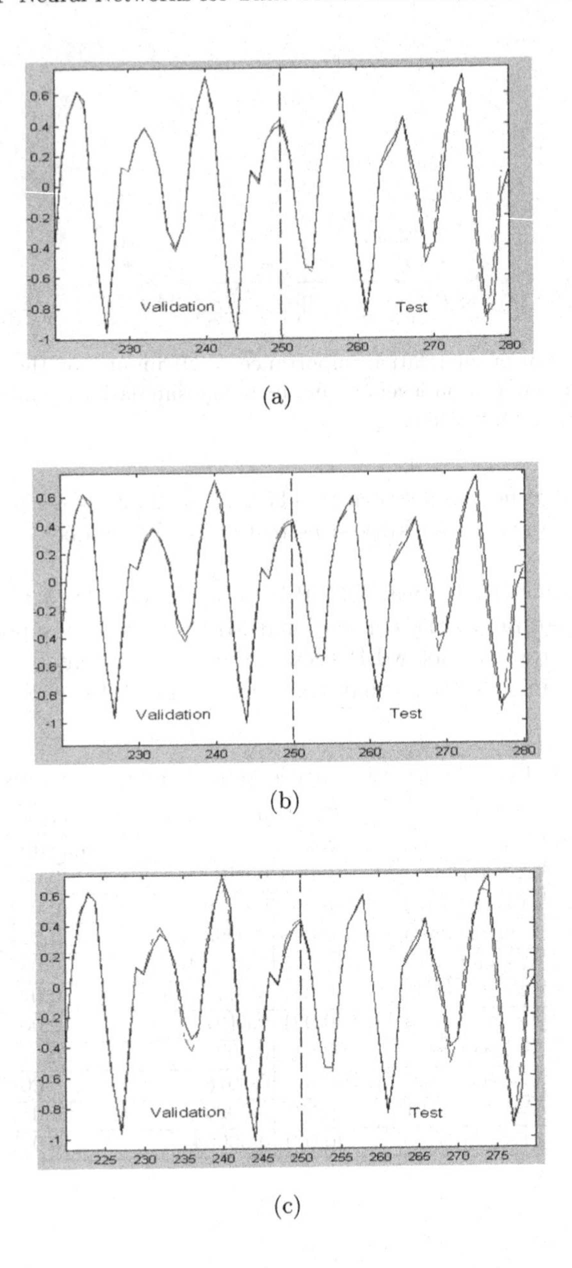

(a)

(b)

(c)

Fig. 2.6. Result for network size 17:2:1 on Mackey-Glass time-series prediction (a) MLP with test and validation NMSE = 0.0028 and 0.0481, respectively. (b) Wavelet MLP with decomposition level one, validation and test NMSE= 0.0068 and 0.0450, respectively. (c) Wavelet MLP (17:2:1) = (20:2:1) -[1,4,5], test and validation NMSE = 0.0126 and 0.0273, respectively. dotted line is the actual data, whereas continuous line is the predicted data.

2.2 Wavelet Packet MLP Neural Networks for Time-series Prediction

2.2.1 Wavelet Packet Multi-layer Perceptron Neural Networks

This section describes the wavelet packet MLP (WP-MLP) and its application to time-series prediction [332]. Instead of decomposing the input using wavelets as in the wavelet MLP studied in the previous section, the WP-MLP is used as a feature extraction method to obtain time-frequency information. The WP-MLP has been successfully applied to classification of biomedical signals, images, and speech.

The WP-MLP consists of three parts, i.e., (1) an input with a tapped delay line with p delays, (2) wavelet packet decomposition, and (3) an MLP. The output $\hat{x}(n+1)$ is the value of the time-series at time $n+1$ and is assumed to be a function of the values of the time-series at p previous time steps.

Wavelet packets were introduced by Coifman *et al.* [61]. Recursive splitting of vector spaces is represented by a binary tree. The wavelet packet transformation is used to produce time-frequency atoms. These atoms provide both time and frequency information with varying resolution throughout the time-frequency plane. The time-frequency atoms can be expanded in a tree-like structure to create an arbitrary tiling, which is useful for signals with complex structures. A tree algorithm [328] can be employed for computing a wavelet packet transform with the wavelet coefficients as filter coefficients.

We decompose the input signal using wavelet packets and then feed into a conventional MLP with k input units, one hidden layer with m sigmoid neurons, and one linear output neuron. The architecture of the WP-MLP is defined as $[p : l(wlet) : h]$, where p is the number of tapped delays, l is the number of decomposition levels, $wlet$ is the type of wavelet packet used, and h is the number of hidden neurons in the MLP.

2.2.2 Weight Initialization with Clustering

Kolen and Pollack [179] showed that a feedforward network with the back-propagation technique can be very sensitive to the initial weight selection. A prototype pattern [74] and the orthogonal least square algorithm [190] may be used to initialize the weights. Usually, the weights and biases are initialized to small random values. If the random initial weights happen to be far from a good solution or they are near a poor local optimum, training may take a long time or get trapped in the local optimum [117]. Proper weight initialization will place the weights close to a good solution, which reduces training time and increases the possibility of reaching a good solution. In this subsection, we describe methods of weight initialization based on clustering algorithms.

Geva *et al.* [117] proposed to initialize the weights by a clustering algorithm based on mean local density (MLD). They showed that this method easily leads to good performance, whereas a random weight initialization leads to a wide variety of different results and many of them were poor. However, we note that the best result from random weight initialization was much better than the result obtained from the MLD initialization method.

Following Geva *et al.* [117], we use a time-frequency event matrix defined by combining the time-frequency patterns and the respective targeted outputs of the training data. Each column is made up of a time-frequency pattern and its respective target. Clustering analysis [159] is the organization of a collection of patterns into clusters based on similarity. Grouping of the time-frequency events is thus revealed. Member events within a cluster are more similar to each other than they are to an event belonging to a different cluster. The number of clusters is then chosen to be the number of neurons in the hidden layer of the WP-MLP. There exist many different methods for clustering. The number of clusters may be chosen before clustering, or may be determined by the clustering algorithm used. Here, the hierarchical tree clustering algorithm [4] and a competitive learning algorithm [228] are used.

The hierarchical tree clustering algorithm consists of the following steps:

1. Compute the Euclidean distance between every pair of events in the data set.
2. Group the events into a binary, hierarchical cluster tree by linking together pairs of events that are in close proximity. As the events are paired into binary clusters, the newly formed clusters are grouped into larger clusters until a hierarchical tree is formed.
3. Cut the hierarchical tree to form a certain number of clusters chosen based on some prior knowledge.

The second clustering method that we use here is competitive learning. Let us consider a competitive learning network consisting of a layer of N_n competitive neurons and a layer of $N+1$ input nodes, N being the dimension of the input pattern, and N_n being the maximum number of clusters to be formed. We have

$$h_k = \sum_{j=1}^{N+1} g_{kj} x_j, \tag{2.13}$$

where g_{kj} is the weight connecting neuron k to all the inputs, h_k is the total input into neuron k, and x_j is element j of the input pattern. If neuron k has the largest total input, it will win the competition and becomes the sole neuron to respond to the input pattern. The input patterns that win the competition in the same neuron are considered to be in the same cluster. The neuron that wins the competition will have its weights adjusted as follows:

$$w_{kj}^{new} = (1 - \alpha(t)) w_{kj}^{old} + \alpha(t) x_j, \tag{2.14}$$

where the learning constant $\alpha(t)$ is a function of time. The other neurons do not adjust their weights. The learning constant will change as follows to ensure that each training pattern has equal statistical importance and is independent of presentation order [331]:

$$\alpha(\tau_k) = \frac{\alpha(1)}{1 + (\tau_k - 1)\alpha(1)}, \tag{2.15}$$

where $\tau_k \geq 1$ is the number of times neuron k has modified its weights, including the current update. At the end of training, those neurons that never had their weights modified are discarded. We thus find the number of clusters.

Once clusters are formed, we proceed to initialize the WP-MLP: the number of hidden neurons of the WP-MLP is chosen to be the same as the number of clusters and the weights of the hidden neurons are assigned to be the centroid of each cluster.

We select three time-series often found in the literature for benchmarking, i.e., the Mackey-Glass delay-differential equation, the yearly sunspot reading, and the laser time-series. Each data set is divided into three portions for training, validation, and testing, respectively. The MLP is trained by the back-propagation algorithm using the Levenberg-Marquadt method [127]. When there is an increase in the validation error, training stops. In order to have a fair comparison, 100 independent simulations are carried out for each weight initialization method and each time-series (weight initialization can be different from trial to trial, since the clustering result can depend on the random initial starting point for a cluster). We record the minimum error, the mean error, and the standard deviation over the 100 runs.

2.2.3 Mackey-Glass Chaotic Time-Series

The values of a and b are chosen as 0.2 and 0.1, respectively. The training data sequences are of length 400, followed by validation and testing data sequences of lengths 100 and 500, respectively.

We use eight neurons in the hidden layer, i.e., there are eight clusters. The competitive learning clustering algorithm is initially given 10 neurons. The experimental results are shown in Table 2.2, Table 2.3, Table 2.4, and Table 2.5.

Table 2.2. Test MSE of the WP-MLP with random initialization on Mackey-Glass chaotic time-series. * stands for wavelet decomposition.

Architecture	Test MSE		
	Mean	Std	Min
[14:1(Db2):8]	1.80×10^{-5}	8.13×10^{-6}	1.02×10^{-5}
[14:2(Db2):8]	3.73×10^{-6}	1.86×10^{-5}	1.25×10^{-6}
[14:3(Db2):8]	1.56×10^{-5}	4.51×10^{-6}	9.65×10^{-6}
*[14:(Db2):8]	1.12×10^{-5}	3.83×10^{-6}	1.49×10^{-6}

Table 2.2 shows the results of prediction errors for the Mackey-Glass time-series using different architectures of the WP-MLP, i.e. the mean, the standard

deviation, and the minimum mean squared error (MSE) of the prediction errors over 100 simulations. Tables 2.3 and 2.4 show the results of prediction errors for networks initialized by hierarchical tree and competitive learning clustering algorithms, respectively.

Table 2.3. Test MSE of the WP-MLP initialized with the hierarchical tree clustering algorithm on Mackey-Glass chaotic time-series. * stands for wavelet decomposition.

Architecture	Test MSE		
	Mean	Std	Min.
[14:1(Db2):8]	5.65×10^{-6}	1.42×10^{-5}	2.16×10^{-6}
[14:2(Db2):8]	2.75×10^{-6}	3.24×10^{-7}	2.31×10^{-6}
[14:3(Db2):8]	3.45×10^{-6}	1.65×10^{-6}	1.53×10^{-6}
*[14:(Db2):8]	2.81×10^{-5}	5.54×10^{-5}	1.38×10^{-5}

Table 2.4. Test MSE of the WP-MLP initialized with the competitive learning clustering algorithm on Mackey-Glass chaotic time-series. * stands for wavelet decomposition.

Architecture	Test MSE		
	Mean	Std	Min.
[14:1(Db2):10]	2.33×10^{-5}	4.30×10^{-5}	3.54×10^{-7}
[14:2(Db2):10]	1.22×10^{-5}	1.21×10^{-5}	2.42×10^{-7}
[14:3(Db2):10]	2.96×10^{-5}	3.52×10^{-5}	9.82×10^{-7}
*[14:3(Db2):10]	2.34×10^{-5}	2.24×10^{-5}	7.86×10^{-6}

The hierarchical tree clustering algorithm provides a consistent performance for the various network architectures except for the network with wavelet decomposition. The competitive learning clustering algorithm leads to the lowest minimum MSE with various network architectures.

Table 2.5 shows that the WP-MLP provides superior prediction performance compared to many others' work, including the discrete-time backpropagation neural network [84], the infinite impulse response (IIR) neural network [211], a network with non-linear preprocessing [56], and the time delay neural network with global feedback [134].

2.2.4 Sunspot and Laser Time-Series

Sunspots are large blotches on the sun that are often larger in diameter than the earth. The yearly average of sunspot areas has been recorded since 1700.

Table 2.5. Comparisons between the WP-MLP results (with random weight initialization WP-MLP/Rm, and weight initializations based on the hierarchical tree WP-MLP/HT and competitive learning WP-MLP/CL clustering algorithms) and other results on the Mackey-Glass time-series found in the literature, i.e., the discrete-time back-propagation neural network (DTB), the infinite impulse response (IIR) neural network, a network with non-linear preprocessing (MLP), and the time delay neural network (TDNN) with global feedback (TDNNGF).

Architecture	Test NMSE		
	Mean	Std	Min.
WP-MLP/Rm	1.86×10^{-5}	1.25×10^{-6}	3.73×10^{-6}
WP-MLP/HT	3.45×10^{-6}	1.65×10^{-6}	1.53×10^{-6}
WP-MLP/CL	1.22×10^{-5}	1.21×10^{-5}	2.42×10^{-7}
DTB [20]	–	–	1×10^{-5}
MLP [22]	–	–	1.05×10^{-5}
IIR [21]	–	–	5.11×10^{-5}
TDNN [23]	–	–	2.8×10^{-4}
TDNNGF [23]	–	–	6.4×10^{-4}

The sunspots of the years 1700 to 1920 are chosen to be the training set, 1921 to 1955 as the validation set, while the test set is taken from 1956 to 1979.

'Set A' in the Santa Fe competition [340] is a clean physics laboratory experiment on a Lorentz-like chaotic behavior in an NH3 far-infrared laser. This time-series includes 1100 samples of amplitude fluctuations in the far-infrared laser, approximately described by three coupled non-linear ordinary differential equations. Samples 1 through 900 are chosen to be the training data, followed by validation and testing data sequences of lengths 100 and 100, respectively.

Our results are comparable to those obtained by Scalenet [117]; however, Scalenet may be regarded as a committee of subnetworks and we expect that a committee of multiple WP-MLPs should improve the performance of a single WP-MLP further.

2.2.5 Conclusion

In this section, we have demonstrated that the WP-MLP can be extremely useful for time-series prediction. We used hierarchical tree and competitive learning clustering algorithms to initialize the WP-MLP. Usually the network weights are initialized to small random values that may be far from a good solution. Simulations show that the weights of the WP-MLP must be initialized to improve performance.

2.3 Cost-Sensitive MLP

In many real-world problems, such as financial analysis and medical diagnosis, different errors in prediction usually lead to different costs. For example, the cost of making an error in predicting the future of a million-dollar investment is higher compared to that of a thousand-dollar investment. The cost of misdiagnosing a person with serious disease as healthy is much greater than misdiagnosing a healthy person as ill. Cost-sensitive neural networks (CSNNs) address these important issues [19]. Cost-sensitive *classification trees* have been studied by Turney [319] and Ting [312].

2.3.1 Standard Back-propagation

The total input to an artificial neuron is

$$h = \sum_{j=1}^{N_I} w_j V_j \ , \tag{2.16}$$

where N_I is the number of inputs to the neuron (dimension of the input vector), $\{w_1, w_2, ..., w_{N_I}\}$ are the weights or synapses of the neuron, and $\{V_1, V_2, ..., V_{N_I}\}$ are the individual inputs to the neuron either from other neurons or external sources of input.

The neuron then determines its output a according to

$$a = f(h + b) \ , \tag{2.17}$$

where b is the bias of the neuron and f is usually a non-linear function, which will be specified later.

Let us consider a layer of N_H neurons. All neurons receive an input vector (pattern) $\{\xi_1, \xi_2, ..., \xi_{N_I}\}$. Neuron j in this layer has weights $\{w_{j1}, w_{j2}, ..., w_{jN_I}\}$. Hence, the total input to neuron j is

$$h_j = \sum_{k=1}^{N_I} w_{jk} \xi_k \ , \tag{2.18}$$

which produces output

$$V_j = g(h_j) = g(\sum_{k=1}^{N_I} w_{jk} \xi_k) \ , \tag{2.19}$$

where

$$g(x) = f(x + b) \ . \tag{2.20}$$

Now let us connect a second layer of N_O neurons on top of this first layer of N_H neurons to form a feedforward neural network. The weight connecting neuron i ($i = 1, 2, ..., N_O$) in the second layer and neuron j ($j = 1, 2, ..., N_H$) in the first layer is W_{ij}. Hence, neuron i in the second layer receives a total input

$$h_i = \sum_{j=1}^{N_H} W_{ij} V_j = \sum_{j=1}^{N_H} W_{ij} g(\sum_{k=1}^{N_I} w_{jk}\xi_k) , \tag{2.21}$$

and produces output

$$\begin{aligned} O_i &= g(h_i) = g(\sum_{j=1}^{N_H} W_{ij} V_j) \\ &= g(\sum_{j=1}^{N_H} W_{ij} g(\sum_{k=1}^{N_I} w_{jk}\xi_k)) . \end{aligned} \tag{2.22}$$

Suppose that for an input pattern $\boldsymbol{\xi}^\mu$ used during training, the desired output pattern is $\boldsymbol{\zeta}^\mu$. We shall use superscript μ to denote the training pattern μ, where $\mu = 1, 2, ..., N_p$ and N_p is the number of input-output training pairs.

The objective of training is to minimize the error between the actual output $\mathbf{O}^\mu$ and the desired output $\boldsymbol{\zeta}^\mu$. A commonly used error measure or *cost function* is

$$\epsilon = \frac{1}{2} \sum_{\mu,i} [\zeta_i^\mu - O_i^\mu]^2 , \tag{2.23}$$

Or, by substituting Eq. (2.22) into Eq. (2.23), we have

$$\epsilon = \frac{1}{2} \sum_{\mu,i} [\zeta_i^\mu - g(\sum_j W_{ij} g(\sum_k w_{jk}\xi_k^\mu))]^2 . \tag{2.24}$$

Back-propagation uses a gradient descent algorithm to learn the weights. For the hidden-to-output connections we have

$$\begin{aligned} \Delta W_{ij} &= -\eta \frac{\partial \epsilon}{\partial W_{ij}} \\ &= \eta \sum_\mu [\zeta_i^\mu - O_i^\mu] g'(h_i^\mu) V_j^\mu \\ &= \eta \sum_\mu \delta_i^\mu V_j^\mu , \end{aligned} \tag{2.25}$$

where η is called the learning rate and we have defined

$$\delta_i^\mu = g'(h_i^\mu)[\zeta_i^\mu - O_i^\mu]. \tag{2.26}$$

For the input-to-hidden connections, we have

$$\begin{aligned}
\Delta w_{jk} &= -\eta \frac{\partial \epsilon}{\partial w_{jk}} \\
&= -\eta \frac{\partial \epsilon}{\partial V_j^\mu} \frac{\partial V_j^\mu}{\partial w_{jk}} \\
&= \eta \sum_{\mu,i} [\zeta_i^\mu - O_i^\mu] g'(h_i^\mu) W_{ij} g'(h_j^\mu) \xi_k^\mu \\
&= \eta \sum_{\mu,i} \delta_i^\mu W_{ij} g'(h_j^\mu) \xi_k^\mu \\
&= \eta \sum_{\mu} \delta_j^\mu \xi_k^\mu \,,
\end{aligned} \tag{2.27}$$

where we have defined

$$\delta_j^\mu = g'(h_j^\mu) \sum_i \delta_i^\mu W_{ij} \tag{2.28}$$

We see that Eq. (2.25) and Eq. (2.27) have exactly the same form, only with different definitions of the δ's. In general, for a feedforward neural network with an arbitrary number of layers, suppose that layer p receives input from layer q, which can be either a hidden layer or the external input. Then, the gradient descent learning rule for layer p can always be written as follows:

$$\Delta w_{pq} = \eta \sum_{\mu} \delta_p^\mu V_q^\mu , \tag{2.29}$$

where δ_p^μ represents the error at the output of layer p and V_q^μ is the input to layer p from layer q. If the layer concerned is the final (or top) layer of the network, δ is given by Eq. (2.26), which represents the error between the desired and the actual outputs. If the layer concerned is one of the hidden layers, δ needs to be calculated with some propagating rule, such as Eq. (2.28). The most popular non-linear function for a neuron is the sigmoid function.

2.3.2 Cost-sensitive Back-propagation

In conventional back-propagation, errors made with respect to different patterns are assumed to be the same, as shown in the cost function given by Eq. (2.23). We now write a cost-sensitive cost function as follows [22]:

$$\epsilon = \frac{1}{2} \sum_{\mu,i} \lambda^\mu [\zeta_i^\mu - O_i^\mu]^2 \,, \tag{2.30}$$

where λ^μ is a cost-dependent factor. The standard back-propagation situation (2.23) is recovered if we let

$$\lambda^\mu = 1 \quad \text{for all } \mu. \tag{2.31}$$

With Eq. (2.30), we can easily generalize the standard back-propagation (SBP) algorithm to cost-sensitive situations. By going through the same derivations as above, we can show that the cost-sensitive cost function given by Eq. (2.30) is minimized by the same back-propagation algorithm, but with Eq. (2.29) modified as follows:

$$\Delta w_{pq} = \eta \sum_{\mu} \lambda^{\mu} \delta_p^{\mu} V_q^{\mu}. \tag{2.32}$$

Other quantities such as the δ's and V's are calculated in the same ways as in the standard back-propagation case.

In an iterative implementation, i.e., all weights are updated after each training pattern is presented, the above cost-sensitive back-propagation (CSBP) can be realized by simply replacing the learning rate η in the standard back-propagation case by a cost-sensitive learning rate $\lambda^{\mu}\eta$ for each training pattern μ.

In the CSBP, 'more important' pattern classes with larger cost factors (λ) have larger learning rates compared to the 'less important' classes with smaller cost factors. Let us consider a case with only two pattern classes. Suppose that $\lambda^{(1)} = 3\lambda^{(2)} = 3$, or making an error in classifying a pattern of class 1 is three times as costly as making an error in classifying a pattern of class 2. The CSBP requires that the learning rate for class 1 is 3η if the learning rate for class 2 is η. This is roughly equivalent to the case where we use the SBP, i.e., the same learning rate for all classes, and present each training pattern in class 2 only once to the network, but present each training pattern in class 1 *three* times.

Suppose that there are a total of N_p input-output pattern pairs for training and there are N_{cl} classes (kinds) of patterns. In particular, there are N_k training patterns for class k, where $k = 1, 2, ..., N_{cl}$. In this book, we assume that there are an equal number of training patterns for each class, i.e.,

$$N_k = N_o \tag{2.33}$$

for all $k = 1, 2, ..., N_{cl}$.

Hence,

$$\sum_{k=1}^{N_{cl}} N_k = N_p = N_{cl} N_o. \tag{2.34}$$

Since, in the SBP cost function (2.23), the sum of all coefficients in front of $[\zeta_i^{\mu} - O_i^{\mu}]^2$ is

$$\sum_{\mu} 1 = N_p \ , \tag{2.35}$$

it is reasonable to require the same condition satisfied in the CSBP. This can be achieved by choosing

$$\lambda^{\mu} = N_{cl} \frac{C_{k(\mu)}}{\sum_{k=1}^{N_{cl}} C_k} \ . \tag{2.36}$$

The SBP is recovered if all C's are the same (thus all λ's are 1).

2.3.3 Experimental Results

Assume that two classes of patterns are uniformly distributed in two intersecting circles in a plane. We randomly generate 600 training patterns (characterized by their coordinates in the plane) for each of the two classes. These patterns are used to train the neural network by setting different cost functions. The neural network has one input layer of two neurons, one hidden layer of three neurons with sigmoid transfer functions, and one output layer of one neuron with a sigmoid transfer function. Three different cost-factor settings are used in the simulation study. The different cost factors are set as follows:

- *Case 1.* $C_1 = C_2 = 0.5$: this is corresponding to the standard back-propagation (SBP) algorithm.
- *Case 2.* $C_1 = 0.2$, $C_2 = 0.8$: this sets a higher cost for class 2 than class 1.
- *Case 3.* $C_1 = 0.8$, $C_2 = 0.2$: this sets a higher cost for class 1 than class 2.

After the network is trained, we randomly generate another 600 test patterns for each class to test the neural network for the recognition rate.

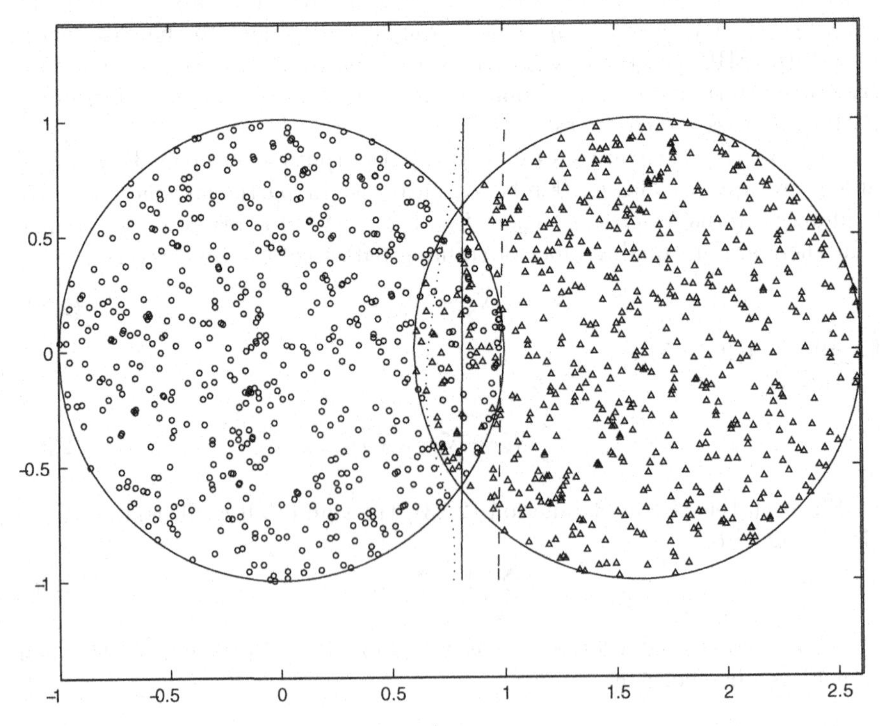

Fig. 2.7. Two classes are represented by triangles and circles, respectively.

The results are shown in Fig. 2.7, where we have computed the decision boundary of the neural network for different cases. The solid line shows the decision line for the standard back-propagation, the dotted line shows the decision line for *Case 2*, and the dashed line shows the decision line for *Case 3*.

The correct recognition rates for the three neural networks using the same set of test patterns are given in Table 2.6.

Table 2.6. Recognition rates for test patterns.

	Class 1	Class 2	Total
Case 1 $C_1 = C_2 = 0.5$	561/600 (93.5%)	571/600 (95.2%)	1132/1200 (94.3%)
Case 2 $C_1 = 0.2, C_2 = 0.8$	542/600 (90.3%)	594/600 (99.0%)	1136/1200 (94.7%)
Case 3 $C_1 = 0.8, C_1 = 0.2$	599/600 (99.8%)	533/600 (88.8%)	1132/1200 (94.3%)

As we can see from Fig. 2.7 and Table 2.6, the recognition rates for the 'important' classes (with higher cost) are higher than the recognition rates for the 'less important' classes. The decision boundaries in Fig. 2.7 also clearly reveal this result.

2.4 Summary

In this chapter, we first used a wavelet MLP, consisting of a wavelet decomposition layer and a conventional MLP, for time-series prediction. We analyzed the relative importance among the input wavelets. After less important wavelets are eliminated, the modified wavelet MLP network provides a more consistent and stable network that is evident in its low mean and standard deviation for NMSE. This is in contrast to the conventional MLP network that has a large performance swing and is sensitive to weight initialization.

However, the wavelet MLP without input elimination did not show significance improvement over the conventional MLP. It is suspected that different signals have different time-frequency compositions. Thus, the decomposition level, type of wavelet, or decomposition type may vary significantly with signal. Therefore, more work is required to equip the network with the ability to adapt to different signals without human intervention.

We also describe a cost-sensitive MLP, in which errors in prediction are biased towards 'important' classes based on the fact that different prediction errors on different classes usually lead to different costs. Experiments show that the recognition rates for the 'important' classes (with higher cost) are higher than the recognition rates for the 'less important' classes.

3

Fuzzy Neural Networks for Bioinformatics

3.1 Introduction

Here we concentrate on the technology of fuzzy logic in intelligent data processing [20][92][239][347] because it provides a highly intuitive and appealing presentation to the end user.

Microarray data analysis is one of those attractive fields of data mining. With the help of gene expressions obtained from microarray technology, heterogeneous cancers can be classified into appropriate subtypes [274]. Recently, different kinds of machine learning and statistical methods [5][34][75][173][306] have been used to analyze gene expression data.

To evaluate the effectiveness of these cancer classification methods, two criteria may be used, i.e., the classification accuracy and the number of genes used by the classifier. For a cancer classifier, the fewer the genes used, the lower the computational burden. A reduced number of genes can significantly increase the classification accuracy because of the reduction or the absence of irrelevant genes acting as "noise" for the classifier. Perhaps more importantly, once a smaller subset of genes are identified as relevant to a particular cancer, it helps biomedical researchers focus on these genes that contribute to the development of the cancer. Therefore, finding the smallest gene subsets that can ensure highly reliable classification results becomes a problem of both theoretical and practical importance.

3.2 Fuzzy Logic

3.2.1 Fuzzy Systems

Fuzzy logic, the logic based upon which fuzzy systems operate, is much closer in spirit to human thinking and natural language than conventional digital logic. Basically, it provides an effective means of capturing the approximate and inexact nature of the real-world knowledge.

Classical logic is referred to as bivalent. Statements are said to be either true or false [182]. This is Aristotle's legacy. He stated that the intersection between set 'A' and set 'NOT-A' was null, an empty set. This was considered to be philosophically correct for over 2000 years. It is also interesting to note that two centuries earlier Buddha held a very different world view. Rather than a cleanly cut perspective of a black-and-white world, he saw a world filled with contradictions. Buddha stated that a rose could be to a certain degree completely red, but also at the same time it was to a degree not red, i.e., the rose can be red and not red at the same time [182]. This is in clear contradiction with an Aristotle's viewpoint.

Fuzzy logic was invented by Lotfi Zadeh in 1964. According to Zadeh [354], the essential characteristics of fuzzy logic are:

- In fuzzy logic, exact reasoning is viewed as a limiting case of approximate reasoning.
- In fuzzy logic, everything is to a matter of degree.
- Any logic system can be fuzzified.
- In fuzzy logic, knowledge is interpreted as a collection of elastic or equivalent, fuzzy constraints on a collection of variables.
- Inference is viewed as a process of propagation of elastic constraints.

Fuzzy logic states that everything is to a matter of degree and fuzzy sets are properties (e.g., low, medium, high) whose elements belong to the sets only in a degree. One example for the fuzzy set *High Speed* is as shown in Fig. 3.1.

The degree of belonging is defined by the value of a membership function (*MF*), which has values between 0 and 1. Such a technique clearly provides a way of representing uncertainties in a mathematical model. The most popular membership functions are the triangular functions, Gaussian functions, bell-shaped functions, and trapezoidal functions. An illustration for bell-shaped membership functions is shown in Fig. 3.2. It is defined as:

$$\mu(x) = \frac{1}{1 + [(\frac{x-c}{a})^2]^b}. \tag{3.1}$$

We define a fuzzy rule as *if* x *then* y, where x (the condition side) is a conjunction in which each clause specifies an input variable and one of the membership functions associated with it, and y (the conclusion side) specifies an output variable membership function. There may be at most one clause for each input variable. For example,

IF input1 = high and input2 = low THEN output = medium.

Fuzzy systems are rule-based expert systems, which comprise a set of fuzzy rules also known as linguistic rules in the form of 'IF–THEN' [160]. Fuzzy

S = Set of High Speed

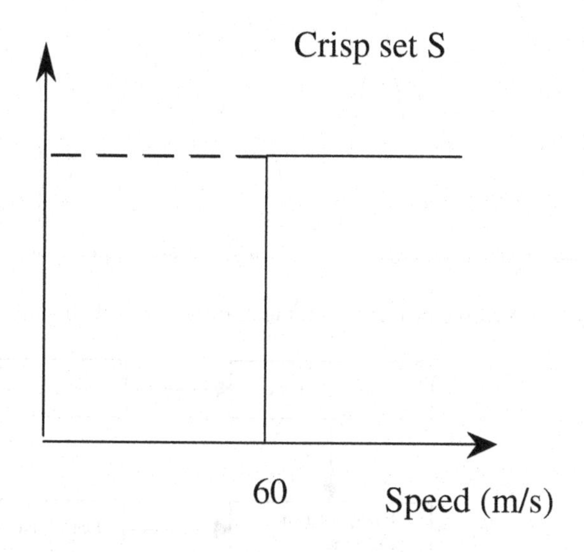

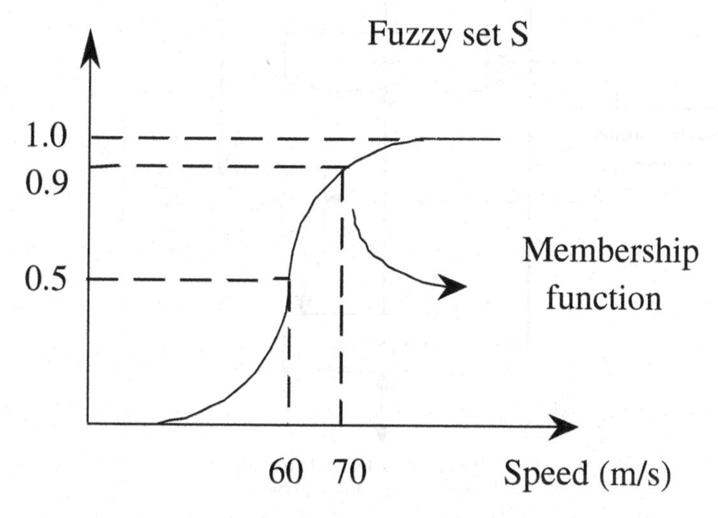

Fig. 3.1. An illustration for a fuzzy set as opposed to a crisp set: high speed

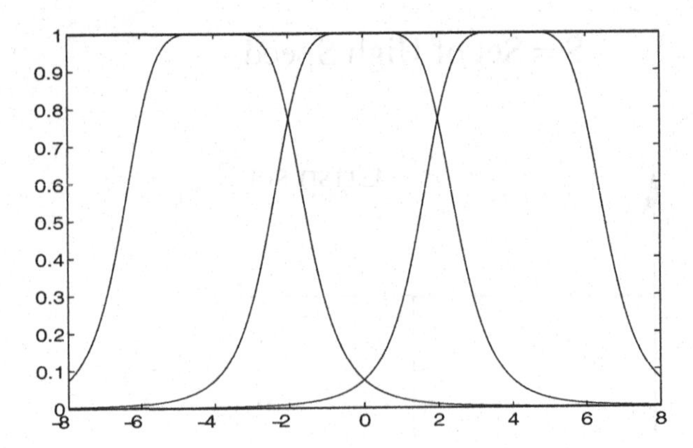

Fig. 3.2. An example of fuzzy membership functions: bell-shaped.

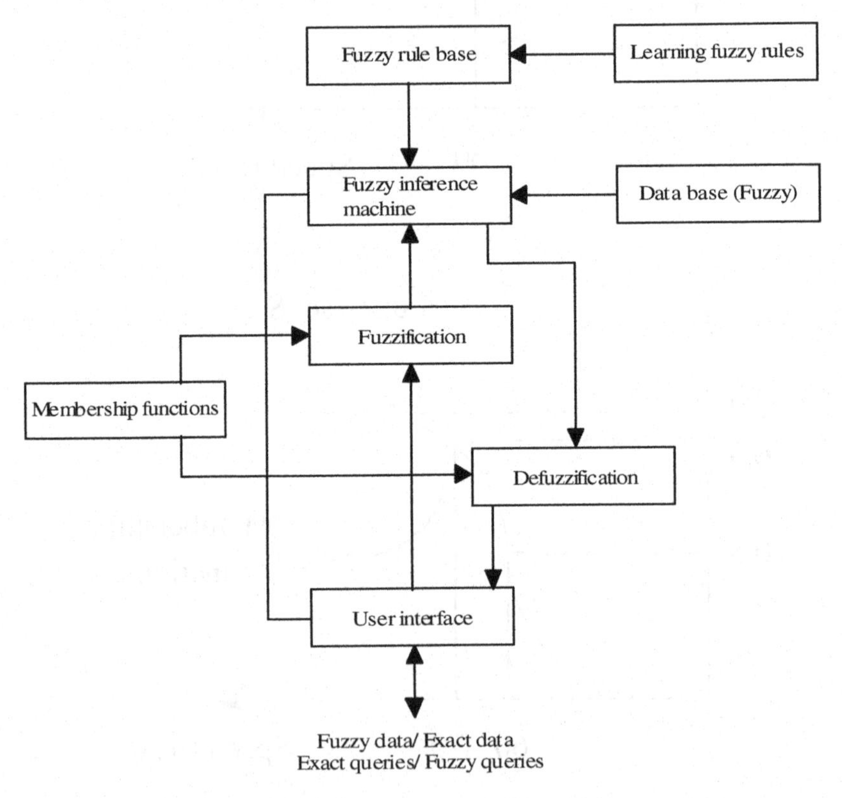

Fig. 3.3. Illustration for a general rule-based fuzzy system.

systems are also known as fuzzy models [295], fuzzy associative memories (FAMs) [181], or fuzzy controllers [188] when used in control applications. A block diagram for a general fuzzy system is shown in Fig. 3.3.

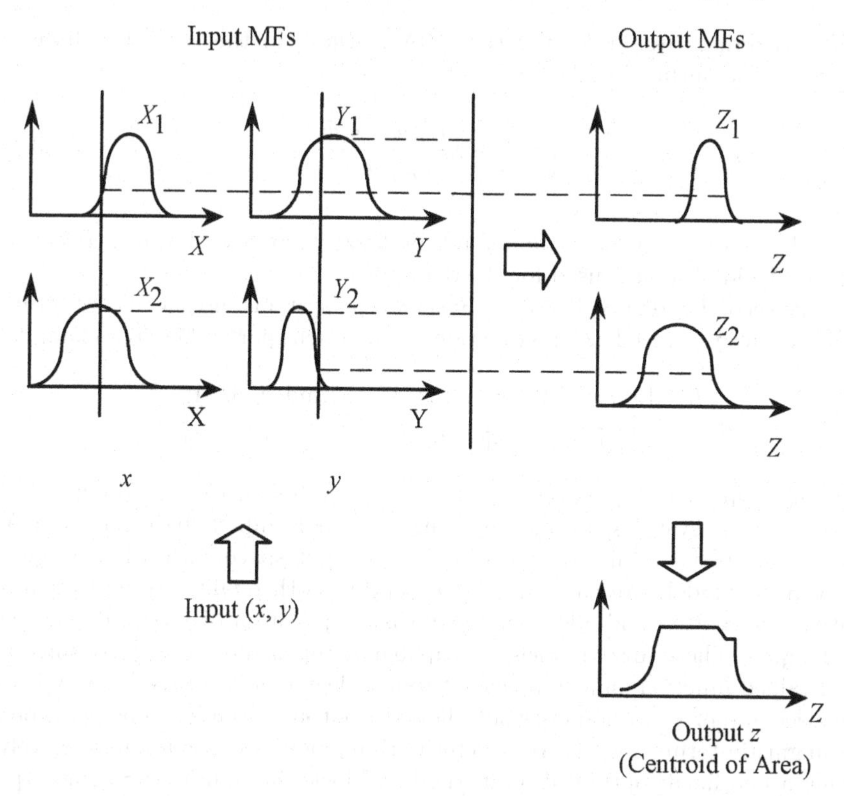

Fig. 3.4. An illustration for a Mamdani-type fuzzy system ('MF' stands for 'membership function').

Fuzzy systems can be broadly categorized into two families. The first includes linguistic models based on collections of IF–THEN rules, whose antecedents and consequents utilize fuzzy values. This family of fuzzy systems uses fuzzy reasoning and the system behavior can be described in natural terms. The *Mamdani* model falls in this group [206]. The knowledge is represented as:

$$R^i : \text{IF } x_1 \text{ is } A_1^i \text{ AND } x_2 \text{ is } A_2^i \dots \text{ AND } x_n \text{ is } A_m^i \text{ THEN } y^i \text{ is } B^i, \quad (3.2)$$

where R^i $(i = 1, 2, \dots, l)$ denotes the ith fuzzy rule, x_j $(j = 1, 2, \dots, n)$ is the input, y^i is the output of the fuzzy rule R^i, and $A_i^1, A_i^2, \dots, A_n^i$, B^i $(i = 1, 2, \dots, l)$ are fuzzy membership functions usually associated with those linguistic terms.

The overall fuzzy output is derived by a 'maximum' operation to the qualified fuzzy outputs (each of them is equal to the minimum of the firing strengths and the output membership function of each rule). Various schemes have been proposed to choose the final crisp output based on the overall fuzzy output. Some of them are centroid of area (COA), mean of maximum (MOM), maximum criterion, etc. [188]. For the COA method, in the case of a discrete universe, the output is [188]:

$$y = \frac{\sum_{i=1}^{u} \mu_y(w_i) \times w_i}{\sum_{i=1}^{u} \mu_y(w_i)}. \tag{3.3}$$

A block diagram for a Mamdani-type fuzzy system is shown in Fig. 3.4. The centroid of area is used for the defuzzifier.

The second category, based on *Sugeno*-type systems, uses a rule structure that has fuzzy antecedent and functional consequent parts [299]. For example,

$$\begin{aligned} R^i : \text{IF } x_1 \text{ is } A_1^i \text{ and } x_2 \text{ is } A_2^i \cdots \text{ and } x_n \text{ is } A_m^i \\ \text{THEN } y^i = a_0^i + a_1^i \times x_1 + \cdots + a_n^i \times x_n. \end{aligned} \tag{3.4}$$

This approach approximates a non-linear system with a combination of several linear systems, by decomposing the whole input space into several partial fuzzy spaces and representing each output space with a linear equation. Such models are capable of representing both qualitative and quantitative information and allow relatively easy application of powerful learning techniques. These fuzzy models are capable of approximating any continuous real-valued function on a compact set to any degree of accuracy. This type of knowledge representation does not allow the output variables to be described in linguistic terms and the parameter optimization is carried out iteratively using a non-linear optimization method. A block diagram for a Sugeno-type fuzzy system is shown in Fig. 3.5.

Experiments demonstrate that the model has the following advantages:

- computational efficiency,
- working well with linear techniques,
- working well with optimization and adaptive techniques,
- continuity of the output surface,
- adapting to mathematical analysis.

It should be highlighted that if no input variables are considered, (3.4) is the same as (3.2), with B^i as a singleton. For convenience, we call this fuzzy model a simplified Sugeno-type model.

Both types of fuzzy models have been extensively used in both system modelling and control purposes. Through the use of linguistic labels and membership functions, a fuzzy IF–THEN rule can easily capture the spirit of "rules of thumb" frequently used by human beings [160]. However, there is a tradeoff between readability and precision [218]. If one is interested in a more precise solution, then he will have to give up some linguistic interpretability. In this book, a simplified Sugeno-type model is used.

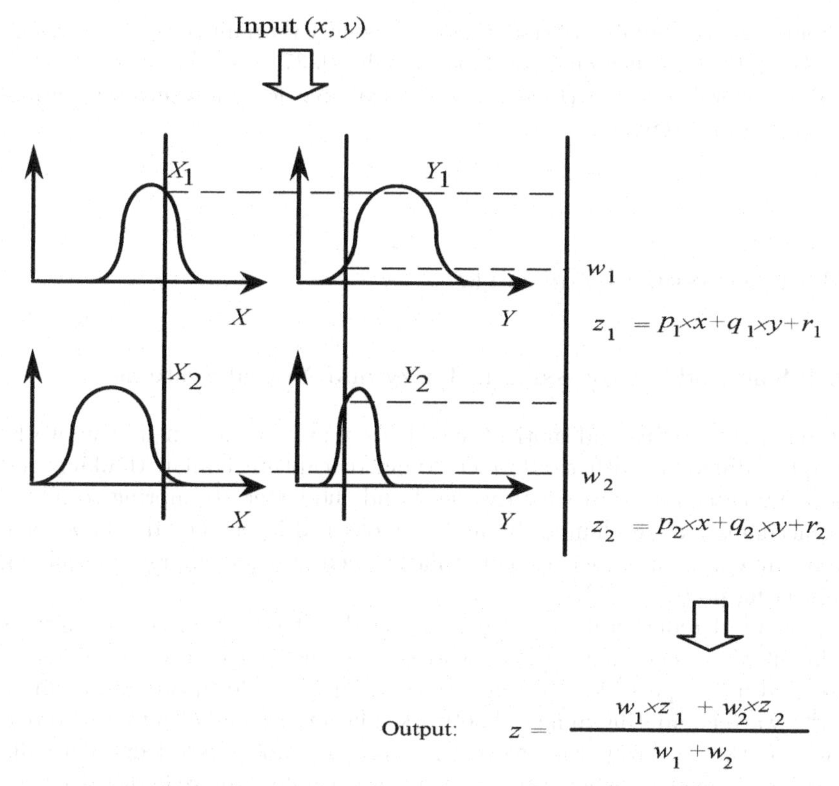

Fig. 3.5. An illustration for a Sugeno-type fuzzy system.

3.2.2 Issues in Fuzzy Systems

From the brief overview of fuzzy systems, we can see that fuzzy systems offer a simple but efficient approach for domain experts to design systems for modelling. In other words, designing a fuzzy system is a subjective approach which is adopted to express domain knowledge of experts. However, domain experts do not structure their decision makings in any formal ways. As a result, the process of transferring expert knowledge into a usable knowledge base is time consuming and non-trivial [188]. Moreover, depending on human experience may result in some severe problems, because even for human experts, their knowledge is often incomplete and episodic rather than systematic. The following questions still remain as open issues.

- Structure identification of fuzzy systems, such as determination of the partition of the input space, the number of membership functions, and the number of rules, may be difficult [296].
- Problems with multi-input and multi-output (MIMO) systems, especially high dimensionality, are often encountered in implementation of fuzzy sys-

tems, and the number of rules increases rapidly for multiple input systems. This is the so-called curse of dimensionality [337].
- Few general and analytical tools are available for analyzing the performance of a fuzzy system.

3.3 Fuzzy Neural Networks

3.3.1 Knowledge Processing in Fuzzy and Neural Systems

Although the origins and motivations of fuzzy systems and neural networks are quite different, with the former trying to capture human thinking and reasoning capability at a cognitive level and the latter attempting to mimic the mechanism of the human brain at a biological level, they do share some important common properties, particularly when they are employed in solving engineering problems [227].

Instead of representing knowledge using IF–THEN localized associations as in fuzzy systems, a neural network stores knowledge through its structure, and more specifically its connection weights and local processing units. Feedforward computing in neural networks plays the same role of forward reasoning as that in fuzzy systems. Fuzzy systems acquire knowledge normally from domain experts, whereas neural networks usually acquire knowledge from samples.

Functionally, a fuzzy system or a neural network can be described as a function approximator, which aims at obtaining an approximation of an unknown mapping $f : \Re^r \rightarrow \Re^s$ from sample patterns drawn from the function f. Theoretical investigations have revealed that both neural networks and fuzzy inference systems are universal approximators [334], i.e., they can approximate any function to any specified accuracy provided that sufficient hidden neurons or fuzzy rules are available.

3.3.2 Integration of Fuzzy Systems with Neural Networks

As stated earlier, fuzzy systems provide an inference morphology that enables approximate human reasoning capabilities to be applied to knowledge-based systems. The theory of fuzzy logic provides a mathematical strength to capture uncertainties associated with human cognitive processes, such as thinking and reasoning. However, a common bottleneck in fuzzy logic is its dependence on the specification of good rules from human experts. Also, there exists no formal framework for the choice of various parameters of fuzzy systems and hence the means of tuning them has become an important subject of fuzzy systems.

Neural networks, on the other hand, offer exciting advantages, such as learning, adaptation, fault tolerance, parallelism, and generalization. They are capable of coping with computational complexity, non-linearity, and uncertainty. In view of this versatility of neural networks, it is believed that they hold great potential as building blocks for a variety of behaviors associated with human cognition. But the main problem in neural networks is that how these networks operate is still not clear to users; this is why neural networks are sometimes called 'black boxes': it can be difficult to interpret the output of a trained network. Hence, some researchers have been studying how to extract knowledge from neural networks [106][108][149][284].

Recently, there has been substantial interest and practice in the synergy of fuzzy systems and neural networks. This has given birth to a rapidly emerging field, fuzzy neural networks (FNNs), which are intended to capture capabilities of both types of systems and overcome drawbacks of each system [125][160][181][227][295][336]. The majority of reported studies of FNNs mainly address one of the following functions:

- Using FNNs to tune fuzzy systems [300].
- Extracting fuzzy rules from given numerical examples [336].

3.4 A Modified Fuzzy Neural Network

In this section, a modified fuzzy neural network (FNN) is described [93]–[99]. This FNN combines the features of initial fuzzy model self-generation, fast input selection, partition validation, parameter optimization and rule-base simplification. The structure, implementation and functionality of this FNN are described in details in the following sections.

3.4.1 The Structure of the Fuzzy Neural Network

The structure of the modified fuzzy neural network is as shown in Fig. 3.6. The network consists of four layers, i.e., the input layer, the input membership function layer, the rule layer, and the output layer.

In databases, data fields are either numerical or categorical. The input membership function layer generates input membership functions for numerical inputs, i.e., numerical values are converted to categorical values. For example, the continuous numerical values of age can be converted to categorical values such as young, middle-aged, and old. Each input node is connected to all membership functions for this input. We use piecewise-linear triangular membership functions (3.5) for computational efficiency, as shown in Fig. 3.7.

$$\mu(x) = \begin{cases} \frac{x-l}{c-l}, & l \le x \le c, \\ \frac{x-r}{c-r}, & c < x \le r, \\ 0, & \text{otherwise.} \end{cases} \tag{3.5}$$

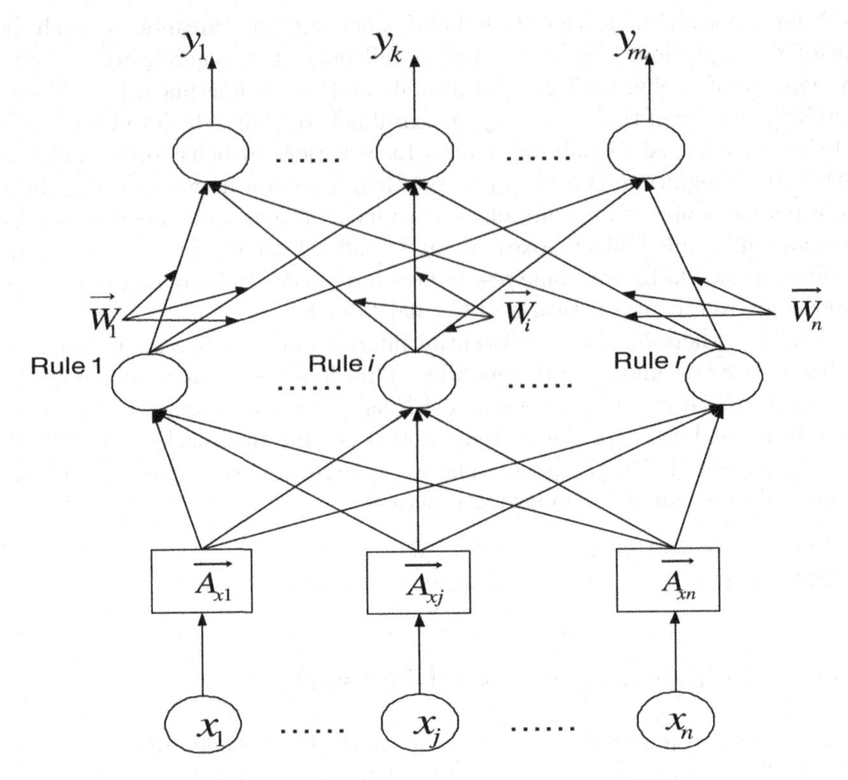

Fig. 3.6. The structure of the modified FNN

The left-most and right-most membership functions are shouldered to cover the whole operating range of each input.

Each rule node is connected to all input membership function nodes and all output nodes for this rule. Each rule node performs a product of its inputs. The input membership functions act as fuzzy weights between the input layer and the rule layer. Links between the rule layer, the output layer, and the input membership functions are adaptive during learning. In the output layer each node receives inputs from all rule nodes connected to this output node.

The flow chart for structure generation and the learning algorithm of the proposed fuzzy neural network are shown in Fig. 3.8. The learning algorithm aims at constructing the fuzzy system by locating the initial membership functions and initial parameters of the consequence linear functions, generating the required fuzzy rules, tuning the membership functions, and updating the consequence parameters and the rule base so that the performance is optimized through the whole set of training data pairs. The detailed explanations are given in the following sections.

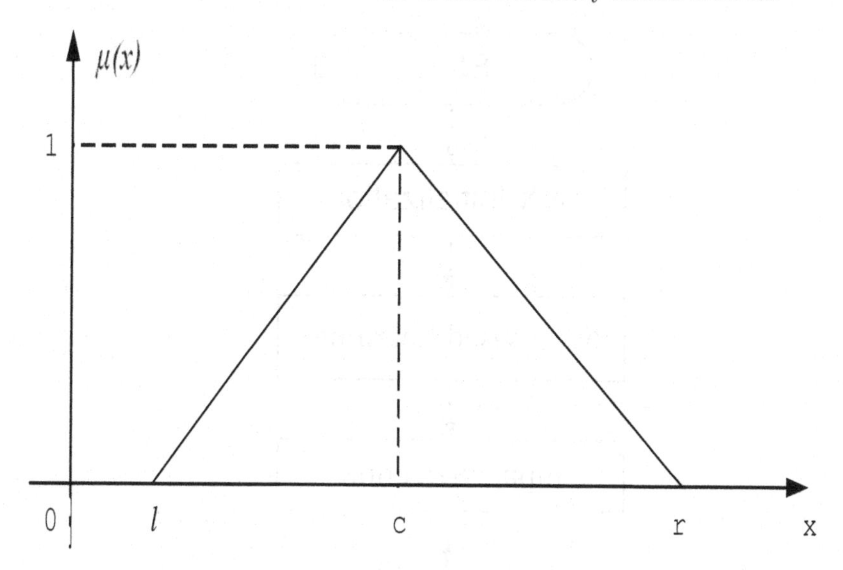

Fig. 3.7. The triangular membership function.

3.4.2 Structure and Parameter Initialization

Firstly, we create n nodes for the input layer and m nodes for the output layer, where n and m are the numbers of the input variables (attributes) and the output variables (classes), respectively. The rule layer is empty, i.e., there are initially no rules in the rule base. Then, we initialize the fuzzy neural network as follows [336]:

Step 1: Divide the input spaces into fuzzy regions
We define the universe of discourse of each input variable as $[x_j^-; x_j^+]$ and then divide each universe of discourse into N regions. Though N could be different for different variables, at the initialization stage we just assume an equal number of divided regions (fuzzy membership functions) for each input variable, which is set to 2. Further modification will be considered in the following stages.

Then, we add two equally spaced input membership functions along the operating range of each input variable. In such a way these membership functions will satisfy ϵ-completeness [188]. The ϵ-completeness means that for a given value of x for one of the input variables in the operating range, we can always find a linguistic label A such that the membership value $\mu_A(x) \geq \epsilon$. In this manner, the fuzzy neural network can provide a smooth transition and sufficient overlapping from one linguistic variable to another. A similar idea can also be found in earlier literature on modular neural networks [50]. As the values of these parameters change by back-propagation training, the triangular-shaped functions vary, thus exhibiting various forms of membership

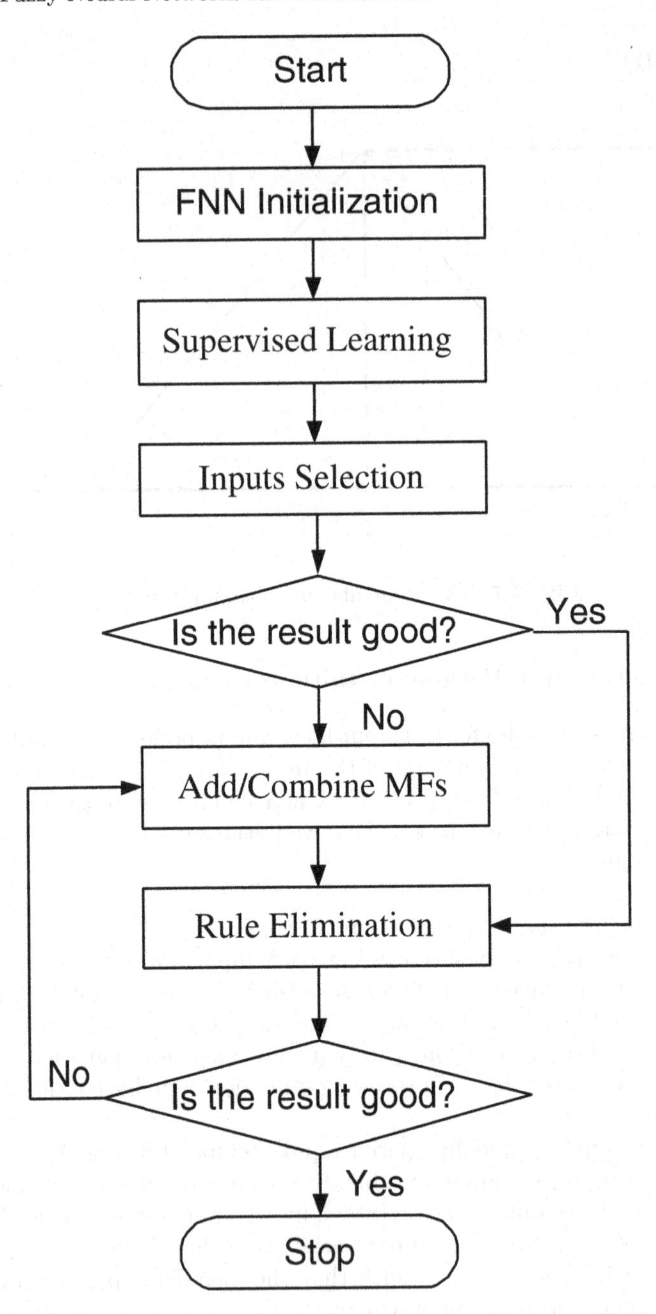

Fig. 3.8. A flow chart of the FNN learning algorithm. (© 2005 IEEE) We thank the IEEE for allowing the reproduction of this figure, first appeared in [59].

functions on a linguistic label. If the ϵ-completeness is not satisfied, there may be no rule applicable for a new training data pair.

The shape of each membership function associated with each region that defines a fuzzy term was assumed triangular, denoted as (l, c, r) for (left bound, center, and right bound). Two special properties of fuzzy terms so defined are: (1) adjacent terms have 1/2 overlap and (2) for the middle terms, the left bound of term i is the center of term $i-1$ and the right bound of term i is the center of term $i+1$.

The minimal and maximum values of each variable are often used to define its universe of discourse. That is, $[x_j^-; x_j^+] = [\min(x_j), \max(x_j)]$. They are also considered to be the center of the left-end term and the right-end term, respectively. That is, $C_{1j} = \min(x_j)$ and $C_{Nj} = \max(x_j)$. Accordingly, the other term center, c_{ij}, can be computed as follows:

$$c_{ij} = \min(x_j) + i \times (\max(x_j) - \min(x_j))/(N-1), \text{where } i = 2, ..., N-1. \quad (3.6)$$

Step 2: Generate fuzzy rules from given training data

In this step, we create the initial rule-base layer using the following form for rule i:

$$\begin{aligned} \text{Rule } i: \quad & \text{IF} \quad x_1 \text{ is } A^i_{x_1} \text{ and } x_2 \text{ is } A^i_{x_2} \cdots \text{ and } x_n \text{ is } A^i_{x_n} \\ & \text{THEN} \;\; y_1 = w^i_1 \;, \;\cdots\;, y_m = w^i_m, \end{aligned} \quad (3.7)$$

where x_j $(j = 1, 2, ..., n)$ and y_l $(l = 1, 2, ..., m)$ are the inputs and the outputs, respectively. w^i_l is a real number. A^i_q $(q = x_1, x_2, ..., x_n)$ is the membership function of the antecedent part of rule i for node q in the input layer.

Firstly, we need to determine the membership degrees of each training data pair for each fuzzy term defined on each region, variable by variable. Secondly, each training data pair is associated with the term having the highest membership degree, again variable by variable. Finally, we can obtain one rule for each training data pair using the term selected in the previous step. The rules so generated are "AND" rules and the antecedents of the IF part of each rule must be met simultaneously in order for the consequent of the rule to occur [336].

The membership value μ_i of the premise of the ith rule, which is also called the degree of the ith rule, is calculated using the product operator

$$\mu_i = A^i_{x1}(x_1) \cdot A^i_{x2}(x_2) \cdots A^i_{xn}(x_n). \quad (3.8)$$

The degree of a rule generated by a training data pair indicates our belief of its usefulness.

The output y_l of the fuzzy inference is obtained using the weighted average:

$$y_l = \frac{\sum_i \mu_i w^i_l}{\sum_i \mu_i} \,. \quad (3.9)$$

Step 3: Pruning the fuzzy rule base

When the number of training data pairs is large, it is quite possible that the same rule could be generated by more than one data pair. These rules are redundant rules. The redundant rules must be removed to maintain the integrity of the rule base. This is achieved by keeping only the rule with the highest degree for each fuzzy region. The one with the highest degree is deemed most useful; therefore, it is kept. Upon this step, the fuzzy rule base is complete.

3.4.3 Parameter Training

The adjusted parameters in the network structure of the FNN can be divided into two categories based on the IF (premise) part and the THEN (consequence) part of the fuzzy rules. In the premise part we need to fine tune the left bounds, centers, and right bounds of triangular membership functions. In the consequence part, the adjusted parameters are the consequence weights.

Once the FNN has been initialized, a gradient-descent-based BP algorithm [344] is employed to adjust the parameters of the fuzzy neural network by using the training data set.

As shown in Fig. 3.6, there are four layers in the FNN. Suppose that the kth layer has $n(k)$ nodes. We can denote the node in the ith position of the kth layer by (k, i), and its node function (or node output) by O_i^k. Since a node output depends on its incoming signals and its parameter set, we have

$$O_i^k = O_i^k(O_1^{k-1}, ..., O_{n(k-1)}^{k-1}, p_1, p_2, p_3, ...), \tag{3.10}$$

where p_1, p_2, $p_3, ...$ are the parameters pertaining to this node. Here we use O_i^k as both the node output and the node function.

Assuming that the given training data set has P entries, we can define the *error measure* (or *energy function*) for the pth ($1 \leq p \leq P$) entry of the training data as the summation of the squared errors:

$$E_p = \sum_{i=1}^{n(4)} (T_{m,p} - O_{m,p}^4)^2, \tag{3.11}$$

where $T_{m,p}$ is the mth component of the pth target output vector and $O_{m,p}^4$ is the mth component of the actual output vector produced by the presentation of the pth input vector. Hence, the overall error measure is $E = \sum_{p=1}^{P} E_p$.

In order to develop a learning procedure that implements gradient descent in E over the parameter space, firstly we have to calculate the *error rate* $\partial E / \partial O$ for p training data and for each output node O. The error rate for the node output at $(4, i)$ can be calculated readily from Eq. (3.11):

$$\frac{\partial E_p}{\partial O_{i,p}^4} = -2 \times (T_{i,p} - O_{i,p}^4). \tag{3.12}$$

For the internal node at (k, i), the error rate can be derived by the chain rule:

$$\frac{\partial E_p}{\partial O_{i,p}^k} = \sum_{m=1}^{n(k+1)} \frac{\partial E_p}{\partial O_{m,p}^{k+1}} \times \frac{\partial O_{m,p}^{k+1}}{\partial O_{i,p}^k}, \tag{3.13}$$

where $1 \leq k \leq 3$. That is, the error rate of an internal node can be expressed as a linear combination of the error rates of the nodes in the next layer. Therefore, for all $1 \leq k \leq 4$ and $1 \leq i \leq n(k)$, we can find $\partial E_p / \partial O_{i,p}^k$ by Eq. (3.12) and Eq. (3.13).

Now, if α is a parameter of the given adaptive network, we have:

$$\frac{\partial E_p}{\partial \alpha} = \sum_{O \in S} \frac{\partial E_p}{\partial O} \times \frac{\partial O}{\partial \alpha}, \tag{3.14}$$

where S is the set of nodes whose outputs depend on α. Then, the derivative of the overall error measure E with respect to α is:

$$\frac{\partial E}{\partial \alpha} = \sum_{p=1}^{P} \frac{\partial E_p}{\partial \alpha}. \tag{3.15}$$

Accordingly, the update formula for the generic parameter α is:

$$\Delta \alpha = -\eta \times \frac{\partial E}{\partial \alpha}, \tag{3.16}$$

in which η is a learning rate, which can be further expressed as:

$$\eta = \frac{k}{\sqrt{\sum_\alpha (\frac{\partial E}{\partial \alpha})^2}}, \tag{3.17}$$

where k is the *step size*, the length of each gradient transition in the parameter space. Usually, we can change the value of k to vary the speed of convergence.

We let the learning rate η vary to improve the speed of convergence, as well as the learning performance (accuracy). We update η according to the following two heuristic rules:

1. If the error measure undergoes five consecutive reductions, increase η by 5%.
2. If the error measure undergoes three consecutive combinations of one increase and one reduction, decrease η by 5%.

Furthermore, due to this dynamical updating strategy, the initial value of η is usually not critical as long as it is not too large.

The learning error ε_l is reduced towards zero or a prespecified small value $\varepsilon_{def} > 0$ as the iteration number k increases:

$$||\varepsilon_{lk}(t)|| \rightarrow 0 \quad \text{or} \quad ||\varepsilon_{lk}(t)|| < \varepsilon_{def} \ , \quad t \in [0, T], \quad \text{as} \quad k \rightarrow \infty \ . \qquad (3.18)$$

3.4.4 Structure Training

In this section, structure learning is used in the modified fuzzy neural network. Firstly, redundant inputs are eliminated. Then, fuzzy partitions are adjusted based on similarity measures. Finally, the fuzzy rule base is modified considering both accuracy and generality.

3.4.5 Input Selection

Real-world classification problems usually involve many attributes and all attributes are not always necessary for the classification task. That is, some attributes may be removable without significant deterioration of the classification ability. If only a few attributes are selected, it is easier to design a fuzzy rule-based classification system with high comprehensibility.

In this step, we evaluate the importance of each input variable based on the initial fuzzy model that incorporates all possible input variables and then eliminate those redundant inputs. The objective of this step is to reduce the input dimensionality of the model without significant loss in accuracy. Elimination of redundant input features may even improve the accuracy.

It is known that the change of system output is contributed by all input variables. The larger the output change caused by a specified input variable, the more important this input may be. The modified fuzzy neural network provides an easy mechanism to test the importance of each input variable without having to generate new models as done in [161]. The basic idea is to let the antecedent clauses associated with a particular input variable i in the rules be assigned a truth value of 1 and then compute the fuzzy output, which is due to the absence of input i. Then, we rank those fuzzy outputs from worst to best, and the worst result indicates that the associated input is the most important.

Furthermore, we have to decide how many inputs should be selected. We start from using the most important input as the only input to the FNN, setting antecedents associated with all other inputs to 1. In the following steps, each time we add one more input according to the ranking order. The subset with the best output result is selected as the input group; other inputs and associated membership functions are deleted.

In Sect. 3.6.2, we will introduce another feature selection method, which is the t-statistics-based approach for gene ranking and selection. Besides the ability of finding the most discriminative features, this method can also be used to handle the correlated feature selection problem [136][215].

3.4.6 Partition Validation

After supervised learning using the back-propagation algorithm, we may find that some of the fuzzy sets in layer two are almost the same. In other words, some term sets of the corresponding universe of discourse have a high degree of similarity. Term sets with a high degree of similarity can be combined into a single term set, that is, they can share a common term node.

In this step, we calculate fuzzy similarities between different fuzzy sets for each input variable. If the degree of overlapping of membership functions is greater than a threshold, we combine those membership functions. We use the following *fuzzy similarity measure* equation [80]:

$$E(A_1, A_2) = \frac{M(A_1 \cap A_2)}{M(A_1 \cup A_2)} \;, \tag{3.19}$$

where $\cap$ and $\cup$ denote the intersection and union of two fuzzy sets A_1 and A_2, respectively. $M(\cdot)$ is the size of a fuzzy set, and $0 \le E(A_1, A_2) \le 1$. The higher $E(A_1, A_2)$, the more similar fuzzy sets A_1 and A_2 are.

From Eq. (3.19), we see that the computation of the similarity of two fuzzy sets requires calculating the size of the intersection and union of two triangular membership functions. For any two fuzzy sets A_1 and A_2, $M(A_1 \cup A_2)$ can be easily derived as follows:

$$M(A_1 \cup A_2) = M(A_1) + M(A_2) - M(A_1 \cap A_2) \tag{3.20}$$

Using the above equation of fuzzy similarity measure, the exact formula for the fuzzy similarity measure of two fuzzy sets with triangular-shaped membership functions, which will be used in our fuzzy neural network, can be derived as below. We consider the similarity measure in six different cases based on the relative positions of membership functions. Figure 3.9 shows the six cases under consideration.

- Case 1: $E(A_1, A_2) = 0$,
- Case 2: $h_{rl} = \frac{L_2 - R_1}{C_1 - C_2 + L_2 - R_1}$,
  $M(A_1 \cap A_2) = \frac{1}{2} \times (R_1 - L_2) \times h_{rl}$,
- Case 3: $h_{rl} = \frac{L_2 - R_1}{C_1 - C_2 + L_2 - R_1}, h_{rr} = \frac{R_2 - L_1}{C_1 - C_2 + R_2 - R_1}$,
  $M(A_1 \cap A_2) = \frac{1}{2} \times (R_1 - L_2) \times h_{rl} - \frac{1}{2} \times (R_1 - R_2) \times h_{rr}$,
- Case 4: $h_{ll} = \frac{L_2 - L_1}{C_1 - C_2 + L_2 - L_1}, h_{rl} = \frac{L_2 - R_1}{C_1 - C_2 + L_2 - R_1}$,
  $M(A_1 \cap A_2) = \frac{1}{2} \times (R_1 - L_2) \times h_{rl} - \frac{1}{2} \times (L_1 - L_2) \times h_{ll}$,
- Case 5: $h_{ll} = \frac{L_2 - L_1}{C_1 - C_2 + L_2 - L_1}, h_{rl} = \frac{L_2 - R_1}{C_1 - C_2 + L_2 - R_1}, h_{rr} = \frac{R_2 - R_1}{C_1 - C_2 + R_2 - R_1}$,
  $M(A_1 \cap A_2) = \frac{1}{2} \times (R_1 - L_2) \times h_{rl} - \frac{1}{2} \times (L_1 - L_2) \times h_{ll} - \frac{1}{2} \times (R_1 - R_2) \times h_{rr}$,
- Case 6: sort L_1, L_2, R_1, R_2 as : $L_a < L_b < R_c < R_d$
  $E(A_1, A_2) = \frac{R_c - L_b}{R_d - L_a}$,
- For Case 2 to Case 5: $M(A_1 \cup A_2) = \frac{1}{2} \times (R_1 + R_2 - L_1 - L_2)$
  $M(A_1 \cup A_2) = M(A_1) + M(A_2) - M(A_1 \cap A_2)$.

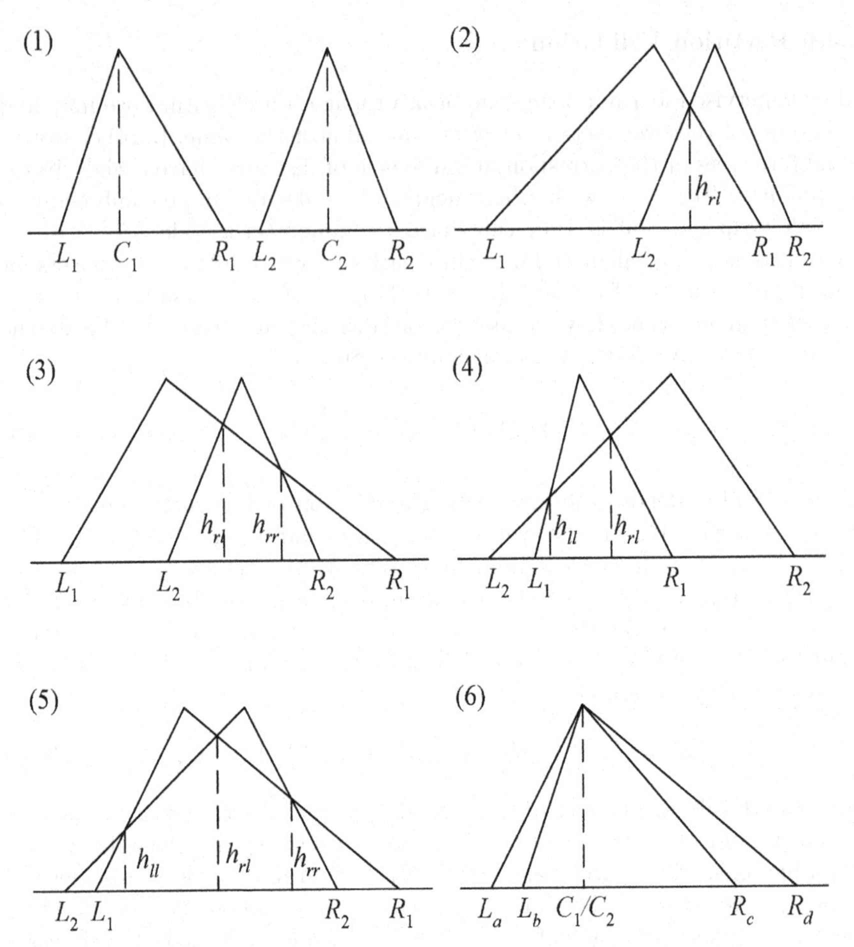

Fig. 3.9. Fuzzy similarity measures of two triangular membership functions at different relative positions.

If an input variable ends up with only one membership function, which means that this input is irrelevant, then we delete the input. We can thus eliminate irrelevant inputs and reduce the size of the rule base.

After this step, if the classification accuracy of the FNN is below the requirement, and the number of rules is less than the specified maximum, we will modify the rule base by following the method introduced in the next section.

3.4.7 Rule Base Modification

In general, when a fuzzy rule-based classification system consists of a large number of specific fuzzy IF–THEN rules, its classification performance on training patterns is very high [157]. Generalization ability on test patterns of such a fuzzy system, however, is not always high due to the overfitting to training patterns. A large number of fuzzy IF–THEN rules also degrade the comprehensibility of fuzzy rule-based classification systems. On the other hand, when a fuzzy rule-based classification system consists of a small number of general fuzzy IF–THEN rules, the comprehensibility is high. Such a compact fuzzy system can avoid the overfitting to training patterns. There are, however, many cases where a large number of fuzzy IF–THEN rules are required for handling complicated non-linear classification problems. That is, when the number of fuzzy IF–THEN rules is too small, fuzzy rule-based classification systems may have poor classification ability on training patterns as well as test patterns. There is a tradeoff between the performance of a fuzzy rule-based classification system and the number of fuzzy IF–THEN rules. That is why we need the rule-base modification procedure.

Firstly, an additional membership function is added for each input at its value at the point of the maximum output error, following Higgins and Goodman [135]. One vertex of the additional membership function is placed at the value at the point of the maximum output error and has the membership value unity; the other two vertices lie at the centers of the two neighboring regions, respectively, and have membership values zero. As the output of the network is not a binary 0 or 1, but a continuous function in the range from 0 to 1, by firstly eliminating the error whose deviation from the target value is the greatest, we can speed up the convergence of the network substantially.

The rules generated above are then evaluated for accuracy and generality. We use a weighting parameter between accuracy and simplicity, which is the compatibility grade (CG) of each fuzzy rule. CG of rule j is calculated by the product operator as:

$$\mu_j(x) = \mu_{j1}(x_1) \times \mu_{j2}(x_2) \times \cdots \times \mu_{jn}(x_n), \tag{3.21}$$

when the system provides correct classification results.

All rules whose compatibility grade (CG) falls below a predefined threshold are deleted. Elimination of rule nodes is rule by rule, i.e., when a rule node is deleted, its associated input membership nodes and links are deleted as well. By varying the CG threshold we are able to specify the degree of rule base compactness. The size of the rule base can thus be kept minimal. If the classification accuracy of the FNN after the elimination of rule nodes is below the prescribed requirement, we will add another rule as described in the previous section; otherwise we will stop the training process.

3.5 Experimental Evaluation Using Synthesized Data Sets

To demonstrate the effectiveness of the modified fuzzy neural network [93], we use 10 classification problems of different complexity defined in [2] on a synthetic database. Experimental results are compared with decision tree construction algorithms C4.5 and C4.5rules [251] and a pruned feedforward crisp neural network (NeuroRule) [200].

3.5.1 Descriptions of the Synthesized Data Sets

There are nine attributes (see Table 3.1 for detailed descriptions) involved for the experiments. Attributes *elevel*, *car*, and *zipcode* are categorical, and all others are non-categorical.

Table 3.1. Descriptions of attributes for the 10 synthesized problems defined in [2].

Attribute	Description	Value
salary	Salary	Uniformly distributed from 20K to 150K
commission	Commission	Salary $\geq 75K$ => commission = 0 else uniformly distributed from 10K to 75K
age	Age	Uniformly distributed from 20 to 80
elevel	Education level	Uniformly chosen from 0 to 4
Car	Make of the car	Uniformly chosen from 1 to 20
zipcode	Zip code of the town	Uniformly chosen from 9 available zipcodes
hvalue	Value of the house	Uniformly distributed from n50K to n150K, where $n \in 0, ..., 9$ depends on zipcode
hyears	Years house owned	Uniformly distributed from 1 to 30
loan	Total loan amount	Uniformly distributed from 0 to 500K

The first classification problem has predicates on the values of only one attribute. The second and third problems have predicates with two attributes, and the fourth to sixth problems have predicates with three attributes. Problems 7 to 9 are linear functions and problem 10 is a non-linear function of attribute values.

Attribute values were randomly generated according to the uniform distribution as in [2]. For each experiment we generated 1000 training and 1000 test data tuples. The target output was 1 if the tuple belongs to class A, and 0 otherwise. We used the random data generator with the same perturbation factor $p = 5\%$ as in [2] to model fuzzy boundaries between classes. The number of data points available for class A (class data distribution) in each problem is as follows: problem 1 = 1321 (66.1% of the total number of data points in this problem), problem 2 = 800 (40.0%), problem 3 = 1111 (55.5%), problem 4 = 626 (31.3%), problem 5 = 532 (26.6%), problem 6 = 560 (28.0%), problem 7 = 980 (49.0%), problem 8 = 1962 (98.1%), problem 9 = 982 (49.1%), and problem 10 = 1739 (87.0%).

The detailed descriptions of the 10 testing functions are as shown below, where $M?N : Q$ is equivalent to the sequential condition, i.e., the expression is equivalent to $(M \bigwedge N) \bigvee (\overline{M} \bigwedge Q)$ [2].

- Function 1
  Class A: $(age < 40) \bigvee ((60 \leq age)$
- Function 2
  Class A: $((age < 40) \bigwedge ((50K \leq salary \leq 100K)) \bigvee ((40 \leq age < 60) \bigwedge (75K \leq salary \geq 125K)) \bigvee ((age \geq 60) \bigwedge (25K \leq salary \leq 75K))$
- Function 3
  Class A: $((age < 40) \bigwedge (elevel \in [0,1])) \bigvee ((40 \leq age < 60) \bigwedge (elevel \in [1,2,3])) \bigvee ((age \geq 60) \bigwedge (elevel \in [2,3,4]))$
- Function 4
  Class A: $((age < 40) \bigwedge (((elevel \in [0,1,2,3,4]?(25K \leq salary \leq 75K)) : (50K \leq salary \leq 100K))) \bigvee ((40 \leq age < 60) \bigwedge (((elevel \in [123])?(50K \leq salary \leq 100K)) : (75K \leq salary \leq 125K))) \bigvee ((age \geq 60) \bigwedge (((elevel \in [2,3,4])?(50K \leq salary \leq 100K)) : (25K \leq salary \leq 75K)))$
- Function 5
  Class A: $((age < 40) \bigwedge (((50K \leq salary \leq 100K)?(100K \leq loan \leq 300K)) : (200K \leq loan \leq 400K))) \bigvee ((40 \leq age < 60) \bigwedge (((75K \leq salary \leq 125K)?(200K \leq loan \leq 400K)) : (300K \leq loan \leq 500K))) \bigvee ((age \geq 60) \bigwedge (((25K \leq salary \leq 75K)?(300K \leq loan \leq 500K)) : (100K \leq loan \leq 300K)))$
- Function 6
  Class A: $((age < 40) \bigwedge (50K \leq (salary + commission) \leq 100K)) \bigvee ((40 \leq age < 60) \bigwedge (75K \leq (salary{+}commission) \leq 125K)) \bigvee ((age \geq 40) \bigwedge (25K \leq (salary + commission) \leq 75K))$
- Function 7
  $disposable = (0.67 \times (salary + commission) - 0.2 \times loan - 20K)$
  Class A: $disposable > 0$
- Function 8
  $disposable = (0.67 \times (salary + commission) - 5000 \times elevel - 20K)$
  Class A: $disposable > 0$
- Function 9
  $disposable = (0.67 \times (salary + commission) - 5000 \times elevel - 0.2 \times loan - 10K)$
  Class A: $disposable > 0$
- Function 10
  $hyears < 20 \Rightarrow equity = 0$
  $hyears \geq 20 \Rightarrow equity = 0.1 \times hvalue \times (hyears - 20)$
  $disposable = (0.67 \times (salary + commission) - 5000 \times elevel - 0.2 \times equity - 10K)$
  Class A: $disposable > 0$

3.5.2 Other Methods for Comparisons

For comparisons with the modified fuzzy neural network, we applied decision tree construction algorithms C4.5 and C4.5rules [251] on the same data sets. C4.5 and C4.5rules release 8 from the pruned trees with default parameters were used. We also compared our results with those of a pruned feedforward crisp neural network (NeuroRule) [200] for the classification problems from [2]. In the next two subsections, we briefly review the principles of C4.5, C4.5rules, and NeuroRule. Experimental results will follow in the subsequent sections.

The Basics of C4.5 and C4.5rules

C4.5 is commonly regarded as a state-of-the-art method for inducing decision trees [78], and C4.5rules transforms C4.5 decision trees into decision rules and manipulates these rules. The heart of the popular and robust C4.5 program is a decision tree inducer. It performs a depth-first, general to specific search for hypotheses by recessively partitioning the data set at each node of the tree.

C4.5 attempts to build a decision tree with a measure of the information gain ratio of each feature and branching on the attribute which returns the maximum information gain ratio. At any point during the search, a chosen attribute is considered to have the highest discriminating ability between the different concepts whose description is being generated. This bias constrains the search space by generating partial hypotheses using a subset of the dimensionality of the problem space. This is a depth-first search in which no alternative strategies are maintained and in which no back-tracking is allowed. Therefore, the final decision tree built, though simple, is not guaranteed to be the simplest possible tree.

C4.5 uses a pruning mechanism to stop tree construction if an attribute is deemed to be irrelevant and should not be branched upon. A χ^2-test for statistical dependency between the attribute and the class label is carried out to test for this irrelevancy. The induced decision tree is finally converted to a set of rules with some pruning and the generation of a default rule.

The C4.5rules program contains three basic steps to produce rules from a decision tree:

1. Traverse a decision tree to obtain a number of conjunctive rules. Each path from the root to a leaf in the tree corresponds to a conjunctive rule with the leaf as its conclusion.
2. Manipulate each condition in each conjunctive rule to see if it can be dropped or can be merged into a similar condition in another rule without more misclassification than expected on the original training examples.
3. If some conjunctive rules are the same after step 2, then keep only one of them.

The final decision rules thus produced are expected to be simpler than the original decision trees but not necessarily more accurate.

The Basics of NeuroRule

NeuroRule is a novel approach proposed by Lu *et al.* [200] that exploits the merits of connectionist methods, such as low classification error rates and robustness to noise. It is also able to obtain rules which are more concise than the rules usually obtained by decision trees.

NeuroRule consists of three steps:

1. Network training: a three-layer feedforward neural network is trained in this step. To facilitate the rule extraction in the third step below, continuous attributes are discretized by dividing their ranges into subintervals. The number of nodes in the input layer corresponds to the dimensionality of subintervals of the input attributes. The number of nodes in the output layer equals the number of classes to be classified. The network starts with the oversized hidden layer and works towards a small number of hidden nodes as well as the fewest number of input nodes. The training phase aims to find the best set of weights that classify input tuples with a sufficient level of accuracy. An initial set of weights are chosen randomly from the interval $[-1, 1]$. Weight updating is carried out by a quasi-Newton algorithm. The training phase is terminated when the norm of the gradient of the error function falls below a prespecified value. A penalty function is used to prevent weights from getting too large or too many with very small values.

2. Network pruning: the pruning phase aims to remove redundant links and nodes without increasing the classification error of the network. The pruning is achieved by removing nodes and links whose weights are below a prespecified threshold. A smaller number of nodes and links left in the network after pruning provide for extracting concise and comprehensive rules that describe the classification function.

3. Rule extraction: this phase extracts classification rules from the pruned network. The rule extraction algorithm first discretizes the activation values of hidden nodes into a manageable number of discrete values without sacrificing the classification accuracy of the network. The algorithm consists of four basic steps:
   (a) A clustering algorithm is applied to find clusters of activation values of the hidden nodes.
   (b) The discretized activation values of the hidden nodes are enumerated and the network outputs are computed. Rules that describe

the network outputs in terms of the discretized hidden unit activation values are generated.

(c) For each hidden node, the input values that lead to it are enumerated and a set of rules is generated to describe the hidden units' discretized values in terms of the inputs.

(d) The two sets of rules obtained in the previous two steps are merged to obtain rules that relate the inputs to the outputs.

The resulting set of rules is expected to be smaller than the one produced by decision trees but with generally more conditions per rule than that of decision trees.

3.5.3 Experimental Results

For each experimental run, the errors for all the groups are summed to obtain the classification error. Table 3.2 shows the overall accuracy on each test data set and the number of rules for all three approaches for the 10 testing problems.

Table 3.2. The overall accuracy on each test data set for the problems from [2], with a compact rule base. The FNN uses the following parameters: learning rate $\eta = 0.1$, maximum number of rules = 20, degree of overlapping of membership functions = 0.8, and maximum number of iterations $k = 200$.

Fnc.	Accuracy (%)			No. of rules		
	NR	C4.5	FNN	NR	C4.5	FNN
1	99.9	100	100	2	3	2
2	98.1	95.0	93.6	7	10	5
3	98.2	100	97.6	7	10	7
4	95.5	97.3	94.3	13	18	8
5	97.2	97.6	97.1	24	15	7
6	90.8	94.3	95.0	13	17	8
7	90.5	93.5	95.7	7	15	13
8	N/A	98.6	99.3	N/A	9	2
9	91.0	92.6	94.1	9	16	15
10	N/A	92.6	96.2	N/A	10	8

Compared to NeuroRule, the modified FNN produces rule bases of less complexity for all problems, except problems 7 and 9. The FNN gives better accuracy for problems 1, 6, 7, and 9, but lower accuracy for the rest. Compared to C4.5rules, the FNN gives less complex rules for all problems. The FNN gives higher accuracy than C4.5rules on problems 6 to 10, the same for problem 1, and lower for the rest. The results for problems 8 and 10 were not reported for the NeuroRule approach [200] (indicated by 'N/A' in Table 3.2).

To have a balanced comparison of all three approaches we have used a *weighted accuracy* (WA), defined as the accuracy A divided by the number of rule conditions C, as our main aim is to achieve a maximum compactness of the rule base while maintaining a high accuracy. The need of the compact rule base is twofold: we should allow for scaling of data mining algorithms to large-dimensional problems, as well as providing the user with easily interpretable rules.

Table 3.3 shows that the FNN performs better than NeuroRule, in terms of WA, on all problems, except problem 7. Compared with C4.5rules, the FNN shows better results for all problems except 10.

Table 3.3. Weighted accuracy on each test data set for NeuroRule (NR), C4.5rules release 8 (C4.5), and the FNN for the problems from [2].

Function	Weighted accuracy (WA)		
	NR	C4.5	FNN
1	20.0	25.0	33.3
2	3.2	4.1	4.8
3	4.7	5.0	6.5
4	1.7	2.2	2.8
5	0.9	2.3	3.1
6	1.5	1.9	2.4
7	4.1	2.8	3.7
8	N/A	5.8	7.1
9	2.9	2.3	3.1
10	N/A	4.2	3.2

In Table 3.2, we attempted to obtain the most compact rule base for the FNN to allow for easy analysis of the rules for very large databases, and subsequently easy decision making in real-world situations.

If accuracy is more important than compactness of the rule base, it is possible to use the FNN with more strict accuracy requirements, i.e., a higher threshold for pruning the rule base, thereby producing more accurate results at the expense of a higher rule base complexity; the classification result based on this requirement is shown in Table 3.4.

Actually, the result in Table 3.4 is a special case where we simply want to achieve higher accuracy. More rules are produced compared with NeuroRule and C4.5rules. For problem 1, compared with the result in Table 3.2, more rules are involved to achieve 100% accuracy, which means that redundant rules are produced as the threshold for rule base compactness decreased. This is only acceptable when accuracy is more important than compactness of the rule base as indicated above. For most cases, the final decision regarding complexity versus accuracy of the rule base is application specific.

Table 3.4. The overall accuracy on each test data set with a higher accuracy for the problems from [2]. The FNN uses the following parameters: learning rate $\eta = 0.1$, maximum number of rules = 200, degree of overlapping of membership functions = 0.8, maximum number of iterations $k = 200$.

Fnc.	Accuracy (%)			No. of rules		
	NR	C4.5	FNN	NR	C4.5	FNN
1	99.9	100	100	2	3	11
2	98.1	95.0	98.5	7	10	25
3	98.2	100	99.3	7	10	36
4	95.5	97.3	98.1	13	18	120
5	97.2	97.6	97.2	24	15	81
6	90.8	94.3	96.5	13	17	50
7	90.5	93.5	97.7	7	15	52
8	N/A	98.6	99.1	N/A	9	15
9	91.0	92.6	95.2	9.0	16	65
10	N/A	92.6	97.7	N/A	10	50

3.5.4 Discussion

Ten different experiments were conducted to test the proposed fuzzy neural network on various data mining problems. The results have shown that our FNN is able to achieve an accuracy comparable to both feedforward crisp neural networks and decision trees, with more compact rules compared to both feedforward neural networks and decision trees for most testing cases. Our FNN is also able to achieve a higher accuracy on most data sets compared to both feedforward neural networks and decision trees if a compact rule base is not required.

With a weighted accuracy, defined earlier as the accuracy divided by the number of rule conditions, our FNN outperforms both feedforward crisp neural networks and decision trees in almost every case.

Compared with the FNN proposed by Frayman and Wang [93], our FNN achieved a similar accuracy and compactness of the rule set as shown in Table 3.5. However, Frayman and Wang's FNN used only fuzzy similarity measure to remove irrelevant inputs; our FNN has a much simpler and effective input selection approach, which combines fuzzy similarity measure and an additional ranking-based selection method.

We also found that incremental online learning of the proposed fuzzy neural network requires less time compared to offline training of feedforward

Table 3.5. A comparison with the FNN in [93], on the overall accuracy of each test data set and the rule base compactness for the problems from [2].

Fnc.	Accuracy (%)		No. of rules	
	FNN in [93]	Modified FNN	FNN in [93]	Modified FNN
1	100	100	2	2
2	94.0	93.6	5	5
3	97.3	97.6	5	7
4	93.3	94.3	9	8
5	93.9	97.1	5	7
6	96.2	95.0	8	8
7	98.7	95.7	11	13
8	99.6	99.3	3	2
9	98.6	94.1	11	15
10	97.7	96.2	8	8

crisp neural networks due to the local updating feature of fuzzy logic, i.e., our FNN only updates the rules applicable to the current data while a feedforward crisp neural network globally updates the network. Our FNN also eliminates the need for a separate phase to extract symbolic rules in both decision trees and feedforward crisp neural networks.

The low accuracy obtained by C4.5rules in the above cases stems from the use by decision trees, such as C4.5rules, of some form of probability estimates for the quality of the rules, most commonly using the relative frequency of data occurrence. These unreliable probability estimates often result in high error rates for classes with less training data. In contrast to decision trees like C4.5rules, the proposed FNN inductive learning is not affected by such a problem. This is due to the fact that our FNN treats all classes with equal importance, while decision trees give preference to more commonly occurring classes and treat classes with less data as less important.

Our proposed FNN can easily make use of existing knowledge. Whenever there exists expert knowledge in the database, it is advantageous to be able to use it. It is relatively difficult to incorporate domain knowledge into crisp neural networks [109][314], while it is easy to incorporate domain knowledge into the proposed FNN.

3.6 Classifying Cancer from Microarray Data

3.6.1 DNA Microarrays

DNA microarrays are a new biotechnology which allows biological researchers to monitor thousands of genes simultaneously [274]. Before the appearance of microarray technology, one traditional molecular biology experiment usually works on only one gene, which makes it difficult to have a "whole picture" of an entire genome. With the help of DNA microarrays, it becomes possible to monitor, analyze, and compare the expression profiles of thousands of genes simultaneously.

DNA microarrays have been used in various fields such as gene discovery, disease diagnosis, and drug discovery. From the end of the last century cancer classification using gene expression profiles has attracted great attention in both the biological field and the engineering field. Compared with traditional tumor diagnostic methods based mainly on the morphological appearance of the tumor, the method using gene expression profiles is more objective, accurate, and reliable. More importantly, some tumors (for example lymphomas) have subtypes which have very similar appearance and are very hard to be classified through traditional methods. It has been proved that gene expression has good capability to clarify these previous muddy problems [5].

Developing accurate and efficient markers based on gene expression data thus becomes a problem that draws attention from both biological and engineering fields. Recent approaches to this problem include artificial neural networks [173], support vector machines [34], k-nearest neighbors [304], nearest shrunken centroids [306], and so on.

It is both useful and challenging to find the important genes that contribute most to the reliable classification of cancers and provide proper algorithms to make a correct prediction based on the expression profiles of those genes. Such work will benefit the early diagnosis of cancers and will help doctors to choose proper clinical treatment. Furthermore, it also helps researchers to find the relationship between those kinds of cancers and the important genes.

DNA microarrays, which are also called gene chips or DNA chips, are valuable tools in areas of research that require the identification or quantization of many specific DNA sequences in complex nucleic acid samples. Microarrays are, in principle and practice, extensions of hybridization-based methods which have been used for decades to identify and quantify nucleic acids in biological samples [87].

On a microarray, samples of interest are labelled and allowed to hybridize to the array; after a sufficient time for hybridization and following appropriate washing steps, an image of the array is acquired and the representation of individual nucleic acid species in the sample is reflected by the amount of hybridization to complementary DNAs immobilized in known positions on the array.

The idea of using ordered arrays of DNAs to perform parallel hybridization studies is not in itself new; arrays on porous membranes have been in use for years [290]. However, many parallel advances have occurred to transform these rather clumsy membranes into much more useful and efficient methods for performing parallel genetic analysis [87]. First, large-scale sequencing projects have produced information and resources that make it possible to

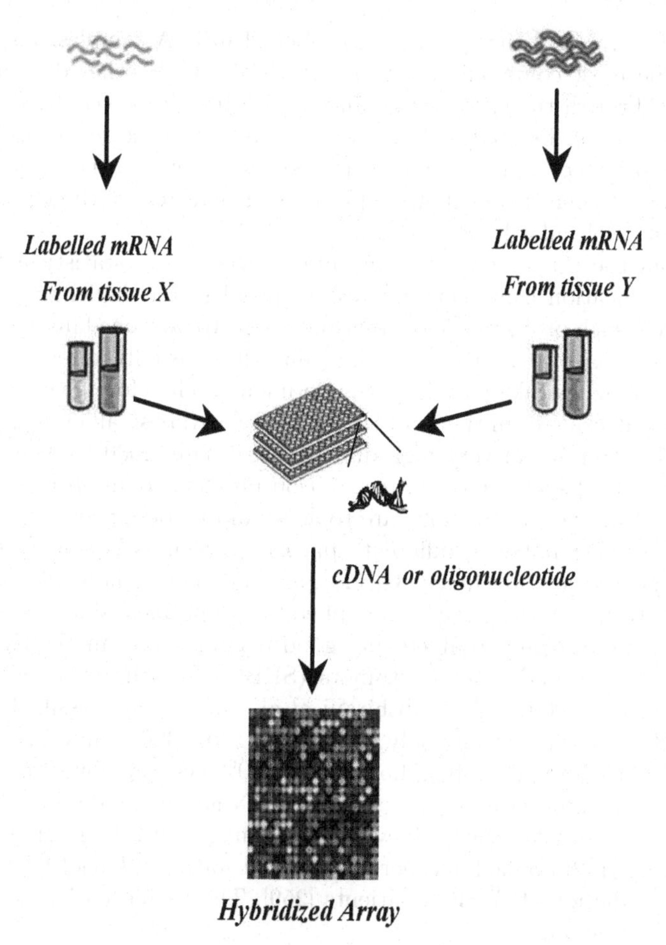

Fig. 3.10. The process of making microarrays.

assemble collections of DNAs that correspond to all, or a large fraction of, the genes in many organisms from bacteria to humans. Second, technical advances have made it possible to generate arrays with very high densities of DNAs, allowing for tens of thousands of genes to be represented in areas smaller than standard glass microscope slides. Finally, advances in fluorescent labelling of nucleic acids and fluorescent detection have made the use of these arrays simpler, safer, and more accurate.

An illustration for the process of making a microarray chip is as shown in Fig. 3.10. The microarray chip is made by spotting a large number of cDNAs onto a small glass slide. Then, the cDNAs hybridize with reference mRNAs. In such experiments, the data for each gene (spot) consist of two fluorescence intensity measurements, (R, G), representing the expression level of the gene in the red (Cy5) and the green (Cy3) labelled mRNA samples, respectively (besides the most commonly used dyes, Cy3 and Cy5, other dyes may also be used). Through the expression value ratio R/G, the molecular differences of those genes can be analyzed. To some degree, gene expression is just like the "fingerprint" of genes; although it does not contain a large quantity of information of genes as a 'photograph' does, it can reflect the characters of genes at the molecular level.

From the point of view of machine learning and statistical learning, cancer classification using gene expression profiles is a challenging problem. For a typical such problem, there are usually tens to over one hundred samples (experiments) in one data set. At the same time, usually several to tens of thousands of genes take part in the experiments. Therefore, such a problem becomes a pattern recognition problem with a relatively small number (usually 20-150) of patterns and very high dimension (usually 2000—10000) of those patterns. To lead such a pattern recognition problem to an optimal solution, appropriate algorithms and software tools should be designed.

In fact, a number of different approaches such as k-nearest neighbors [352], support vector machines [34][114], artificial neural networks [173], and some statistical methods have been applied to this problem since 1995. Among these approaches, some obtained very good results. For example, Khan *et al.* classified small round blue cell tumors (SRBCTs) with 100% accuracy by using 96 genes in 2001 [173]. Tibshirani *et al.* successfully classified SRBCTs with 100% accuracy by using only 43 genes in 2002 [306]. They also classified three different subtypes of lymphoma with 100% accuracy by using 47 genes [306]. In 2002, Ando *et al.* [7] developed a FNN for the prediction of survival of DLBCL patients [5]; they achieved 93% accuracy with four genes. In 2003, Honda *et al.* [142] applied another FNN-based software, GeneFIS, for cancer outcome prediction of DLBCL patients [259]. They achieved 73.4% accuracy by using four genes.

However, we still need to further improve the present algorithms, because usually two criteria may be used to evaluate the effectiveness of these cancer classification methods, i.e., the classification accuracy and the number of genes used by the classifier. For a cancer classifier, the fewer the genes used, the lower the computational burden. A reduced number of genes can significantly increase the classification accuracy, because of the reduction or the absence of irrelevant genes acting as 'noise' for the classifier. Perhaps more importantly, once a smaller subset of genes is identified as relevant to a particular cancer, it helps biomedical researchers focus on these genes that contribute to the development of the cancer. Therefore, finding the smallest gene subsets that can ensure highly reliable classification results becomes a problem of both theoretical and practical importance.

3.6.2 Gene Selection

Gene expression data are very high-dimensional data. The dimension of the input patterns is determined by the number of genes used. In a typical microarray experiment, usually several thousands (for example 6000) of genes take part. Consequently, the dimension of the patterns is several thousands. However, only a part of the genes contribute to classification; some others even act as 'noise'. Gene selection can eliminate the influence of such 'noise' genes. In the meanwhile, once a smaller subset of genes is identified as relevant to a particular cancer, it helps biomedical researchers focus on these genes that contribute to the development of the cancer. Finally, fewer genes used mean a lower computational burden for the classifier. The process of gene selection is ranking the genes' importance for classification and then picking out the genes with high ranks.

As a critical step for cancer classification, gene selection has been studied intensively in recent years. There are two main approaches; one is principal component analysis (PCA), perhaps the most widely used method; the other is t-statistics, which has been more and more widely accepted. In important papers [5][173][243], PCA was used. The basic idea of PCA is to find the most 'informative' genes that contain the main information of the data set. Another approach is based on a statistical approach, t-test, which is a method to measure the difference between two groups. Thomas *et al.*[304] suggested this method. Tusher *et al.* [320] and Pan [236] also proposed their method based on the t-test, respectively. Besides these two main methods, some other methods have also been used. For example, a method called Markov blanket is recommended by Xing *et al.* [350]. Li *et al.* [192] applied another method, which combined genetic algorithm and k-nearest neighbors. In this book, we chose a t-test based gene-selection method, which achieved a better result compared to other methods.

PCA for Gene Selection

The principal component analysis aims at reducing the input dimension by transforming the input space into a new space described by principal components (PCs). All the PCs are orthogonal and they are ordered according to the absolute value of their eigenvalues. The kth PC is the vector with the kth biggest eigenvalue. By ignoring the vectors with small eigenvalues, the input space's dimension is reduced.

In fact, the PCs indicate the directions with largest variations of input vectors. Because PCA chooses vectors with the largest eigenvalues, it can cover directions in which the largest variations of the vectors happen in the vector space. In the direction determined by the vectors with small eigenvalues, the variations of the vectors are also very small. In a word, PCA intends to capture the most informative directions.

We tested PCA [58][59] as the gene selection scheme in the lymphoma data set [5]. We chose 62 genes from 4026 genes in the data set by using PCA. Then, we ranked those genes according to their eigenvalues (absolute values). Finally, we used our proposed fuzzy neural network to classify the lymphoma data set. At first, we randomly divided the 62 samples into two parts, 31 samples for training and the other 31 samples for testing. We then added the selected 62 genes one by one to the FNN according to their eigenvalue ranks starting with the gene ranked 1. That is, we first used only a single gene that is ranked 1 as the input to the FNN. We trained the FNN with the training data set and subsequently tested the FNN with the test data set. We repeated this process with the first two genes, then three genes, and so on. Fig. 3.11 shows the testing error rate and the training error rate. From the results, it is found that the classifier cannot reach 100% accuracy for both training data set and the testing data set. The best testing accuracy is 92.31% that happened when six or 44 genes were input to the classifier. Our t-test-based classification results will be shown in the next section, which are much better than the PCA approach.

t-statistics-Based Approach

A t-statistics-based method can be used to measure the difference between two groups. For example, two groups, x_1 and x_2 have n_1 and n_2 samples, respectively. $\overline{x}_1$ and $\overline{x}_2$ are the means of x_1 and x_2. A typical t-statistics (t) between group x_1 and group x_2 can be described as follows [77]:

$$t = \frac{\overline{x}_1 - \overline{x}_2}{S_p \cdot \sqrt{\frac{1}{n_1} + \frac{1}{n_2}}}, \tag{3.22}$$

where:

$$S_p^2 = \frac{(n_1 - 1)S_1^2 + (n_2 - 1)S_2^2}{n_1 + n_2 - 2}, \tag{3.23}$$

$$S_i^2 = \frac{\sum_{j=1}^{n_i}(x_{ij} - \overline{x}_i)^2}{n_i - 1} \qquad \text{i=1,2.} \tag{3.24}$$

The absolute value of t can tell us the distinction between the distributions of these two groups. In view of this point, we consider searching for the genes which have the largest distinction between different cancers with the help of t-statistics.

One challenge is that a gene expression data set usually has more than two classes, but t-statistics can only be applied to two groups. Therefore, we consider picking out the genes which show the largest distinction between one specific class and the rest of the other classes. Thus, we define a new factor (TS) for gene selection by modifying the typical t-statistics [59]. The TS of gene i is defined as follows:

$$TS_{ik} = \left|\frac{\overline{x}_{ik} - \overline{x}_{ik-else}}{S_i\sqrt{\frac{1}{n_k} + \frac{1}{n-n_k}}}\right|, \tag{3.25}$$

where:

$$\overline{x}_{ik-else} = \frac{\sum_{j,j \notin C_k} x_j}{n - n_k}, \tag{3.26}$$

$$S_i^2 = \frac{1}{n-K}\sum_k \sum_{j \in C_k} (x_{ij} - \overline{x}_{ik})^2. \tag{3.27}$$

The number of classes is K. For each class C_k, it has n_k samples. The total number of samples is n. The expression value of gene i in sample j is denoted as x_{ij}. The mean expression value in class k for gene i is represented as $\overline{x}_{ik}$ and $\overline{x}_{ik-else}$ is the mean expression value for gene i in all the classes except C_k. The pooled within-class standard deviation for gene i is denoted as S_i. Actually, TS used here is a t-statistics between a specific class and the centroid of all the remaining classes. A gene ranking result of the lymphoma data set using TS is shown in Table 3.6.

From the definition of TS, it is found that each gene has K different TSs. In fact, this enables us to design different gene selection schemes. If we use the maximum of these TSs for a specific gene as the mark to rank genes, we get the genes which are the most capable of classifying a specific class. If we use the sum of these TSs for a specific gene as the mark, we get the genes which have good general classification capability. Here, we use the maximum scheme. That is, the maximum of TS_{ik} is used as the mark to rank genes. Practical experiments in the next section will prove that this scheme works quite well for our FNN classifier and gets results much better than the PCA approach [58].

3.6.3 Experimental Results

In this section, we will analyze four well-known gene expression data sets [58][59], which are the small round blue cell tumors (SRBCTs) [173], the lymphoma data set [5], the leukemia data set [121], and the liver cancer data set [51].

Lymphoma Data Set

This data set [5] can be obtained from the website http://llmpp.nih.gov/lymphoma. Expression profiles of 62 lymphoma samples were produced with a "Lymphochip" containing 17856 cDNA clones. A subset of 4026 clones was selected by the authors for being 'well measured' across the samples. The samples represent the following types of lymphoid malignancies: diffuse large

B-cell lymphoma (DLBCL, 42 samples), follicular lymphoma (FL, nine samples), and chronic lymphocytic leukemia (CLL, 11 samples). In this data set, a small part of the data is missing. A k-nearest neighbor algorithm was applied to fill those missing values [316].

As a first step to select important genes for classification, we ranked the entire 4026 genes according to their t-scores (TSs). Then we picked out the 196 genes with the highest t-scores (Table 3.6). In this book, we named every gene by its importance rank. For example, gene 3 means the gene ranked 3 in Table 3.6. Through its ID in the microarray (for example GENE1622x), the real name of each gene can be found on the website of the lymphoma data set [5]. We then used the proposed fuzzy neural network to classify the lymphoma microarray data set.

We randomly divided the 62 samples into two parts, 31 samples for training and the other 31 samples for testing. We then added the selected 196 genes one by one to the FNN according to their t-score ranks starting with the gene ranked 1 in Table 3.6. That is, we first used only a single gene that is ranked 1 as the input to the FNN. We trained the FNN with the training data set and subsequently tested the FNN with the test data set. We repeated this process with the first two genes in Table 3.6, then three genes, and so on. We found that the FNN performed very well: it can reach 100% accuracy for all the training and testing data sets with only the first six genes in Table 3.6. The training error and the testing error of the classification for the lymphoma data set are shown in Fig. 3.12.

Here the 100% accuracy for the training data set indicated that the FNN fits the training data very closely, whereas the 100% accuracy for the testing data ensures the strong generalization ability of the FNN. As we required 100% accuracy for both training and testing data sets, this approach will not suffer from the overfitting problem.

SRBCT Data Set

The small round blue cell tumors (SRBCTs) data set [173] is a widely referenced data set. This group of highly malignant neoplasms accounts for approximately 10% of all solid tumors to affect children under the age of 15 years, based on incidence. They are generally composed of small round cells that appear blue when stained by conventional histopathological processes. Owing to their morphological similarities, unambiguous clinical diagnosis is difficult. Nevertheless, SRBCTs display highly diverse biological behaviors and therefore early diagnosis is essential in order to select an appropriate therapy.

The expression data set for the SRBCT includes four types of cancers, neuroblastoma (NB), rhabdomyosarcoma (RMS), non-Hodgkin lymphoma (NHL), and the Ewing family of tumors (EWS). There are 23 EWS samples, 8 BL samples, 12 NB samples, and 20 RMS samples in the data set. The expression profiles of 2308 genes are available in the data set.

A similar approach is applied to the SRBCT data set. Firstly, we selected out the 30 genes with the highest t-scores (Table 3.7). Then, we applied the proposed fuzzy neural network as the classifier to the SRBCT microarray data set. As provided in the SRBCT data set, there are 63 samples for training and 20 samples for testing. We found that the proposed FNN can achieve 100% accuracy for both the training and the testing data sets with only the first seven genes in Table 3.7. The plots for the training error and the testing error are shown in Fig. 3.13.

Liver Cancer Data Set

The liver cancer data set [51] is available at the authors' website (http://genome-www.stanford.edu/hcc/). We processed the data set to classify non-tumor liver samples and hepatocellular carcinoma (HCC) samples. After some preprocessing steps [51], the authors gave the expression profiles of 1648 important genes. The data contains 156 samples in total. Among them, 82 are HCC; the other 74 are non-tumor liver samples. In this data set, there are some missing values. We used the k-nearest neighbor method to fill those missing values [316].

In the liver cancer data set, we followed the same steps as before. We found that the proposed FNN can achieve 100% accuracy for both the training and the testing data sets with only 24 genes. The plots for training error and testing error are shown in Fig. 3.14.

Leukemia Data Set

The Leukemia data set [121] can be obtained at http://www-genome.wi.mit.edu/cgi-bin /cancer/publications. The samples in this data set belong to two types of leukemia, acute myeloid leukemia (AML) and acute lymphoblastic leukemia (ALL). Among these samples, 38 of them are used for training; the other 34 independent samples are for testing. The entire leukemia data set contains the expression information of 7129 genes. Different from the cDNA microarray data, the Leukemia data is oligonucleotide microarray data.

Because such expression data are raw data, i.e., the unprocessed data, we need to normalize it to reduce the systemic influence induced by the different distributions of raw data in experiments.

We followed the normalization procedure used in [83]. Three preprocessing steps were applied. (a) thresholding with a floor of 100 and ceiling of 16000; (b) filtering, exclusion of genes with $\max/\min \leq 5$ or $(\max - \min) \leq 500$, max and min refer to the maximum and minimum, respectively; (c) base 10 logarithmic transformation. There are 3571 genes surviving after these three steps. After that, the data were standardized across experiments, i.e., minus the mean and divided by the standard deviation of each experiment.

In the Leukemia data set, we followed the same procedure as before. We found that the proposed FNN can achieve 100% accuracy for both the training and the testing data sets with only 46 genes. The plots for the training error and the testing error are shown in Fig. 3.15.

Discussion

Until now, the best published result on cancer classification with gene expression data is from *the Laboratory for the Statistical Analysis of Microarray Data* in Stanford University (http://www-stat.stanford.edu/~tibs/lab/index.html). In 2002, they obtained 100% classification accuracy in the SRBCT data set with 43 genes [306]. They also obtained 100% classification accuracy in the lymphoma data set in the same year [306]. These results are the best published results. Their statistical method is named the nearest shrunken centroid. The software (PAM) can be obtained from their web site [306].

To make a comparison, we also analyzed the SRBCT and the lymphoma data sets with the nearest shrunken centroid method. The results of the SRBCT data set and the lymphoma data are shown in Fig. 3.16 and Fig. 3.17, respectively. Fig. 3.17 is a little different from the result in [306] because only 58 samples are used in [306]. However, we used all the 62 samples given in Fig. 3.17, which makes the comparison more reasonable.

In 2001, Khan *et al.* [173] successfully classified the SRBCT data set with 100% accuracy by using 96 genes. In their opinion, this is the first application of an artificial neural network (ANN) for cancer classification using gene expression data.

In Table 3.8, we made a comparison of the classification result of our FNN classifier with the nearest shrunken centroid [306] and the ANN classifier proposed by Khan *et al.* [173]. In this table, we found that all the three classifiers can obtain 100% classification accuracy, but the number of genes used for classification shows great differences. Our FNN classifier uses only seven genes, in sharp contrast with 96 genes used by the ANN and 43 genes used in the nearest shrunken centroid.

In Table 3.9, a similar comparison is made for the classification result of the lymphoma data set. We also find that our FNN classifier uses much fewer genes to obtain 100% classification accuracy than the nearest shrunken centroid. Compared with other classifiers, our proposed fuzzy neural network classifier has a high accuracy with a fewer number of genes in most cases.

As we mentioned in the previous part, the number of genes used to classify different cancers is a factor of crucial importance. A smaller number of genes helps researchers directly focus their attention on some specific genes, which perhaps will lead to the discovery of deep reasons for the development of cancers and the discovery of drugs. From this point of view, our FNN shows great superiority over other published classifiers.

3.7 A Fuzzy Neural Network Dealing with the Problem of Small Disjuncts

3.7.1 Introduction

Despite many advantages in data mining approaches, such as classification trees and feedforward neural networks using back-propagation-type learning rules, some aspects require improvements. A notable problem is known as the problem of small disjuncts, where the induced rules that cover a small amount of training cases often have high error rates. The purpose of this section is to show that a dynamically constructed recurrent fuzzy neural network can deal effectively with this problem [96].

. Decision trees, such as C4.5 [249], are generated through symbolic inductive algorithms [30][249] and use the maximum generality bias to achieve a high predictive accuracy on disjuncts that cover a large proportion of training instances (large disjuncts). However, the use of a maximum generality bias often results in high error rates on disjuncts that cover a small number of training instances (small disjuncts). The low accuracy in the latter cases stems from the use by decision trees such as C4.5 of some form of probability estimates (most commonly the relative frequency) to estimate the quality of inductive rules. The unreliable probability estimates from a small number of training instances resulted in *the problem of small disjuncts*, where specific rules often produce high error rates [141][310][341][342].

Some remedies exist for C4.5 to overcome the problem of small disjuncts, for example, modifying the original Bayes-Laplace formula [250], and using a composite learner consisting of C4.5 and an instance-based learning method [3] for small disjuncts [310]. Another possibility is to assign different misclassification costs for different classes that in effect will change the original frequency of data. However, this is not a straightforward task as, for example, the instance-based learning method suffers from a similar problem, termed *the problem of atypicality* [311]. Thus, the problems of small disjuncts and atypicality can be seen as manifestations of an intrinsic problem in learning systems. In this section, we will approach the problem of small disjuncts using a fuzzy neural network that does not use a probability estimate in its rule induction. Thus, it should not depend on a relative frequency of a particular class in the overall data set.

3.7.2 The Structure of the Fuzzy Neural Network Used

The fuzzy network used in this section consists of four layers [93]–[99], i.e., the input, the input membership function, the rule, and the output layers. The input nodes represent input variables consisting of the current inputs and the previous outputs. This recurrent structure provides the possibility to include temporal information, i.e., the network learns dynamic input-output mapping instead of the static mapping in feedforward neural networks.

In the input layer, each input node is connected to all membership function nodes for this input. The input membership functions act as fuzzy weights between the input layer and the rule layer. Piecewise-linear triangular membership functions that correspond to second-order B-splines [35] are used. This type of membership functions is simple to implement and is computationally efficient. The left-most and right-most membership functions are shouldered to cover the whole operating range of the input.

Rule node i represents fuzzy rule $i = 1, 2, ..., r$:

$$\begin{aligned} &\text{IF } x_1 \text{ is } A^i_{x_1} \text{ AND } ... \; x_n \text{ is } A^i_{x_n} \\ &\text{AND } y_1(k-1) \text{ is } A^i_{y_1} \text{ AND } ... \; y_m(k-1) \text{ is } A^i_{y_m} \\ &\text{THEN } y_1 = w^i_1 \;, \; ... \;, \; y_m = w^i_m \;. \end{aligned} \tag{3.28}$$

Here x_j $(j = 1, 2, ..., n)$, and y_l $(l = 1, 2, ..., m)$, are the inputs and the outputs, respectively. w^i_l is a real number. A^i_q $(q = x_1, x_2, ..., x_n, y_1, y_2, ..., y_m)$ is the membership function of the antecedent part of rule i for node q in the input layer, and k is the time.

Each rule node is connected to all input membership function nodes and output nodes for this rule. Links between the rule layer and the output layer, and the input membership functions are adaptive during learning. The membership value μ_i of the premise of the ith rule is calculated as fuzzy AND using the product operator:

$$\begin{aligned} \mu_i = A^i_{x1}(x_1) \cdot A^i_{x2}(x_2) \cdots A^i_{xn}(x_n) \cdot A^i_{y1}(y_1) \\ \cdot A^i_{y2}(y_2) \cdots A^i_{ym}(y_m) \;, \end{aligned} \tag{3.29}$$

where μ_i indicates the degree to which the compound antecedent of the rule is satisfied.

The use of a product operator makes fuzzy inference fully differentiable at any point [229][257]. On the other hand, the use of the truncation (MIN) operator would introduce derivative discontinuities both along the lines parallel to the input axes and along the main and minor diagonals [35]. The relational surface in this case would also have large areas where the system is not sensitive to any changes in input, and the output of the fuzzy system is constant in these regions. Therefore, fuzzy inference that uses truncation operators is not inherently robust; rather, the information lost during their operation produces an undesirable fuzzy relational surface [35].

In addition, the use of the product operator for fuzzy AND produces a smooth output surface, in contrast with the commonly used fuzzy MIN operator. The product operator forms multivariate membership functions. Such membership functions retain more information than when the MIN operator is used. The latter scheme retains only one piece of information, whereas the product operator combines n pieces [35].

In the output layer, each node receives inputs from all rule nodes connected to this output node. The output y_l of the fuzzy inference is obtained using the weighted average (or center of gravity defuzzification):

$$y_l = \frac{\sum_i \mu_i w_l^i}{\sum_i \mu_i} \ . \tag{3.30}$$

The use of a weighted average (following the simplified fuzzy inference) produces a smoother output than the mean of maxima (MOM) defuzzification method and greatly reduces both the computational cost and the storage requirement of the algorithm [35].

The FNN structure generation and learning algorithms are as follows. Initially, $n + m$ nodes for the input layer and m nodes for the output layer are created. Here n and m are the numbers of the input variables (attributes) and the output variables (classes), respectively. Next, two equally spaced input membership functions are added along the operating range of each input variable. In such a way these membership functions satisfy ϵ-completeness, which means that for a given value of x of one of the inputs in the operating range, we can always find a linguistic label A such that the membership value $\mu_A(x) \geq \epsilon$. If the ϵ-completeness is not satisfied there may be no rule applicable for a new data input. The initial rule layer is created using Eq. (3.28).

The network is trained using the following general learning rule:

$$y_l^i(k+1) = y_l^i(k) - \eta \frac{\partial \varepsilon_l}{d y_l^i} \ . \tag{3.31}$$

The learning rules for w_l^i and A_j^i are:

$$w_l^i(k+1) = w_l^i(k) - \eta \frac{\partial \varepsilon_l}{\partial w_l^i} \ , \tag{3.32}$$

for adaptation of the weights between the rule layer and the output layer, and

$$A_q^i(k+1) = A_q^i(k) - \eta \frac{\partial \varepsilon_l}{\partial A_q^i} \ , \tag{3.33}$$

for adaptation of membership functions (fuzzy weights). Here η is the learning rate.

The objective is to minimize an error function:

$$\varepsilon_l = \frac{1}{2}(y_l - y_{dl})^2 \ . \tag{3.34}$$

Here y_l is the current output and y_{dl} is the target output.

The learning rate η is variable: a relatively large learning rate to enhance the learning speed is used initially, i.e., $\eta = 0.01$. Whenever the error

$\varepsilon(t)$ starts increasing, the learning rate is reduced according to the following iterative formula:

$$\eta_{\text{new}} = r_c \, \eta_{\text{old}} \ . \tag{3.35}$$

Here r_c is a coefficient in the range $(0, 1)$. Such a decreasing learning rate algorithm can improve the speed of convergence, resulting in substantial reduction in training time, as well as in improvements in learning performance (accuracy).

The following recursive procedure is employed next. If the degree of overlapping of membership functions is greater than a specified threshold (e.g., 0.9), those membership functions are combined. The following *fuzzy similarity measure* [80] is used:

$$E(A_1, A_2) = \frac{M(A_1 \cap A_2)}{M(A_1 \cup A_2)} \ . \tag{3.36}$$

Here $\cap$ and $\cup$ denote the intersection and union of two fuzzy sets A_1 and A_2, respectively. $M(\cdot)$ is the size of a fuzzy set and $0 \leq E(A_1, A_2) \leq 1$.

If an input variable ends up with only one membership function, which means that this input is irrelevant, delete the input. Irrelevant inputs can thus be eliminated and the size of the rule base can be reduced. Combining the membership functions is also done to eliminate poor membership functions and to replace them with new ones that are likely to perform better.

If the classification accuracy is above or if the number of rule nodes is below the respective prespecified threshold, the algorithm stops. Otherwise, add an additional membership function for all inputs at the point of the maximum output error. By firstly eliminating the errors whose deviation from the target values is the greatest, the output error can be reduced more efficiently and the convergence of the network can be substantially improved.

Since we would like to find a small set of simple and accurate rules, there is a need to achieve a maximum compactness of the rule base while maintaining a high accuracy. The need of the compact rule base is twofold: it should allow for scaling of data mining algorithms to large-dimensional problems, as well as providing the user with easily interpretable rules.

We therefore also have a pruning phase in the algorithm. The generated rules are evaluated for accuracy and generality. A weighting parameter between accuracy and generality, the rule applicability coefficient (weighting of the rules) (WR), is used. It is defined as the product of the number of the rule activation RA by the accuracy of the rule A. All rules whose rule applicability coefficient WR falls below a predefined threshold are deleted. When a rule node is deleted, associated input membership functions and links are deleted as well. By varying the WR threshold the user is able to specify the degree of rule base compactness. In such a way, the size of the rule base can be kept minimal. Thus, effectively both construction and pruning phases are employed in the overall algorithm.

3.7.3 Experimental Results

We tested the FNN method on a car evaluation database from the Machine Learning Repository at the University of California, Irvine [23]. The car evaluation database was derived from a simple hierarchical decision model [25][363]. The reason for using this database is to exemplify the problem of small disjuncts when the data distribution is skewed.

The car evaluation database contains examples with the structural information removed, i.e., directly relates a car to six input attributes: buying, maint, doors, persons, lug-boot, and safety. The database has 1728 instances that completely cover the attribute space, six attributes and four classes. The attribute values are: buying = {v-high, high, med, low}, maint = {v-high, high, med, low}, doors = {2, 3, 4, 5-more}, persons = {2, 4, more}, lug-boot = {small, med, big}, and safety = {low, med, high}. The class distribution is as follows: class unacc = 1210 (70.0%), acc = 384 (22.2%), good = 69 (4.0%), vgood = 65 (3.8%). The data set was randomly split into two equal sets: training and testing, with the same distribution of classes in each set.

For comparison with the FNN approach, we used C4.5Rules [251] to make the results comparable with the rule-based FNN. C4.5Rules release 8 with default parameters was used. The performance of the FNN and C4.5Rules on a car evaluation database is given in Table 3.10.

The FNN gives higher accuracy than C4.5Rules for the whole database as well as per each class, with a less complex rule base for class 'acc'. While both the FNN and C4.5Rules give quite similar accuracy for classes 'unacc' and 'acc', the FNN gives much higher accuracy for classes 'good' and 'vgood'. Note that, the frequency for class 'good' in the database is 4.0%, and for class 'vgood' 3.8%.

3.8 Summary

In this chapter, we described a novel fuzzy neural network (FNN) and demonstrated its applications to data classification problems, especially for cancer classification based on microarray data.

The proposed fuzzy neural network combines the features of initial fuzzy model self-generation, fast input selection, partition validation, parameter optimization, and rule-base simplification.

A small FNN is created from scratch, i.e., there is no need to specify the initial network architecture, initial membership functions, or initial weights. Fuzzy IF-THEN rules are constantly combined and pruned to minimize the size of the network while maintaining accuracy; irrelevant inputs are detected and deleted, and membership functions and network weights are trained with a gradient descent algorithm, i.e., error back-propagation.

Experimental studies of synthesized data sets demonstrate that the proposed fuzzy neural network is able to achieve accuracy comparable to or higher than both a feedforward crisp neural network, i.e., NeuroRule, and a decision tree, i.e., C4.5, with more compact rule bases for most of the data sets used in our experiments.

The FNN has achieved outstanding results for cancer classification based on microarray data. A t-statistics-based approach is used for effective gene selection, which is shown to have better performance than the principal component analysis (PCA) approach. The excellent classification results for the lymphoma data set, the small round blue cell tumors (SRBCTs) data set, the leukemia data set, and the liver Cancer data set are shown. Compared with other published methods, we have used a much fewer number of genes for perfect classification, which will help researchers directly focus their attention on some specific genes and may lead to the discovery of deep reasons for the development of cancers and the discovery of drugs.

In addition, we described a fuzzy neural network dealing with small disjuncts. As can be seen from the results in Table 3.10, the FNN inductive learning (i.e., the error rate) is affected much less by the problem of small disjuncts, in contrast to C4.5Rules. This is because the FNN treats all classes with equal importance, whereas C4.5Rules gives preference to classes with a higher occurrence in the data set. On the other hand, C4.5Rules treats classes with lower support as less important (considering them as noise or exceptions). In reality, however, classes with low occurrence in the data set may not be noise, instead contain essential knowledge that one tries to extract from the database. For example, for the car evaluation database used in our experiments, the aim is not only to find which car is not suitable, but also to find which car is the most suitable. The FNN can offer much better advice to the user on finding cars in 'good' and 'vgood' classes, in comparison to C4.5Rules. Thus, the FNN approach to small disjuncts can be very useful in real-world situations, as the data of interest can often be only a small fraction of the available data.

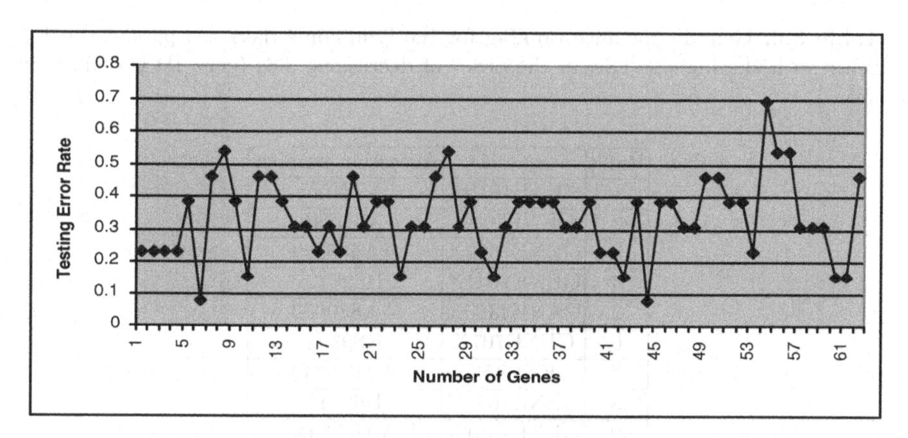

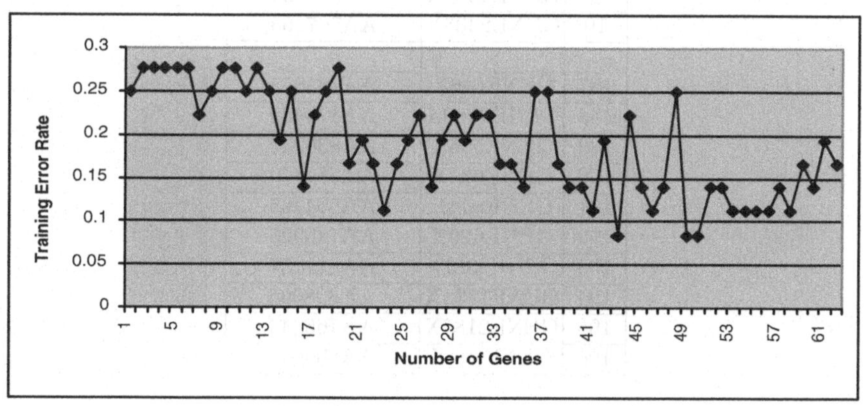

Fig. 3.11. PCA classification results for lymphoma data set.

Table 3.6. Gene importance ranking for the lymphoma data set: part of the 196 genes with the highest TSs, in the order of decreasing TSs (gene ID is defined in [5]).

Rank	Gene ID	Accession number
1	GENE1610X	X72755
2	GENE708X	X65550
3	GENE1622X	AA430369
4	GENE1641X	R62612
5	GENE3320X	AA830983
6	GENE707X	R89392
7	GENE653X	AA258849
8	GENE1636X	R62612
9	GENE2391X	AA937964
10	GENE2403X	AA767265
...	...	...
187	GENE646X	AA765853
188	GENE2180X	AA828464
189	GENE506X	AA836242
190	GENE632X	AA761420
191	GENE844X	AA504465
192	GENE629X	AA740926
193	GENE2381X	AA811187
194	GENE1533X	AA808006
195	GENE2187X	AA765843
196	GENE641X	AA806641

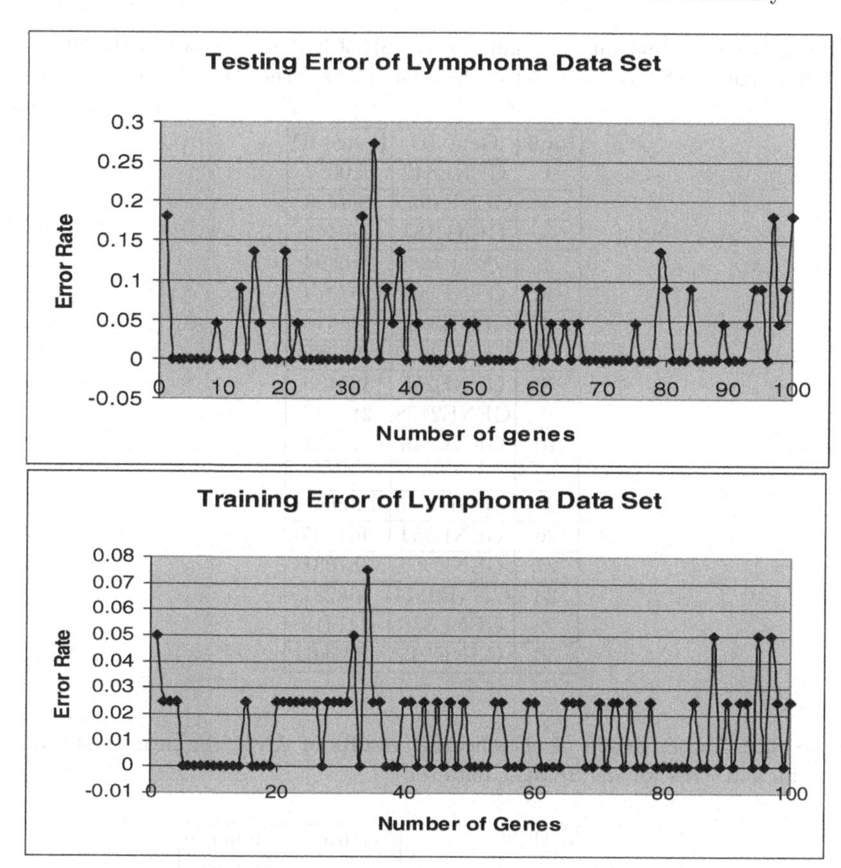

Fig. 3.12. The FNN classification result for the lymphoma data set with 100% training and testing accuracy using only six genes. (© 2005 IEEE) We thank the IEEE for allowing the reproduction of this figure, first appeared in [59].

Table 3.7. Gene importance ranking for the SRBCT data set: part of the 30 genes with the highest TSs, in the order of decreasing TSs (gene ID is defined in [173]).

Rank	Gene ID	Image ID
1	GENE842	810057
2	GENE1955	784224
3	GENE187	296448
4	GENE1389	770394
5	GENE509	207274
6	GENE2046	244618
7	GENE2050	295985
8	GENE255	325182
9	GENE2198	212542
10	GENE246	377461
...	...	...
25	GENE1055	1409509
26	GENE554	461425
27	GENE566	357031
28	GENE2144	308231
29	GENE836	241412
30	GENE545	1435862

Table 3.8. A comparison of classification results of ANN, the nearest shrunken centroid, and FNN for the SRBCT data set.

Method	Accuracy	Number of genes used
ANN	100%	96
Nearest shrunken centroid	100%	43
Fuzzy neural network	100%	7

Table 3.9. A comparison of classification results of the nearest shrunken centroid and FNN for the lymphoma data set.

Method	Accuracy	Number of genes used
Nearest shrunken centroid	100%	48
Fuzzy neural network	100%	6

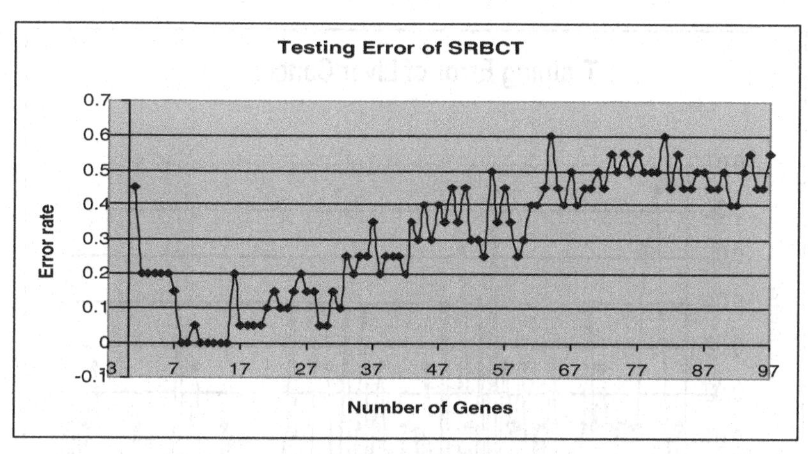

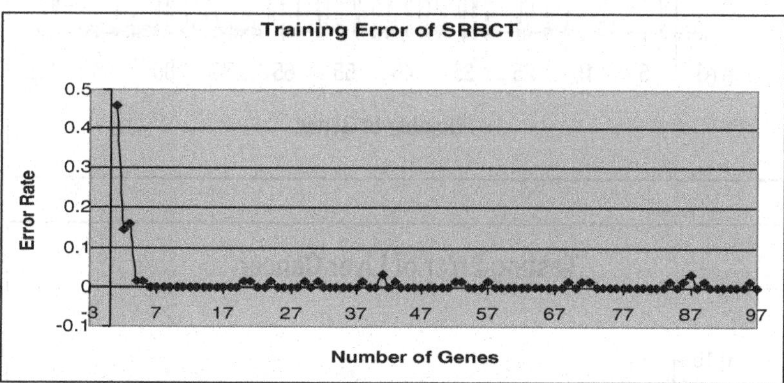

Fig. 3.13. The FNN classification result for the SRBCT data set with 100% training and testing accuracy using only seven genes. (© 2005 IEEE) We thank the IEEE for allowing the reproduction of this figure, first appeared in [59].

Table 3.10. Accuracy of the test data set, the number of rules, the number of conditions for C4.5Rules release 8 (C4), and the FNN (FN) for the car evaluation database.

Class	Accur. (%)		Rules		Cond.	
	C4	FN	C4	FN	C4	FN
unacc	98.0	98.4	9	10	20	24
acc	72.0	79.7	25	14	97	58
good	44.1	82.9	6	8	24	28
vgood	54.5	75.0	6	7	25	31
Overall	88.4	92.7	46	39	166	141

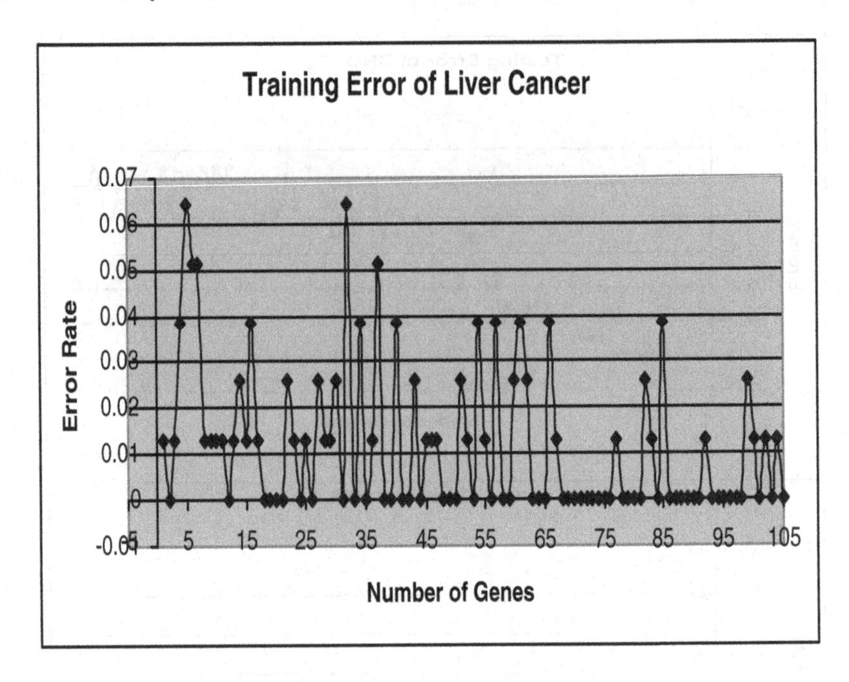

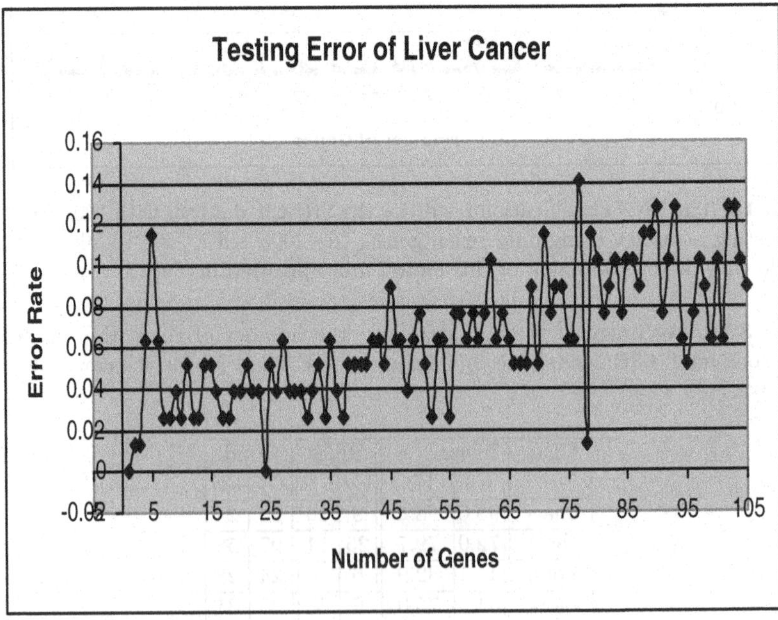

Fig. 3.14. FNN classification result for liver cancer data set, with 100% training and testing accuracy using only 24 genes. (© 2005 IEEE) We thank the IEEE for allowing the reproduction of this figure, first appeared in [59].

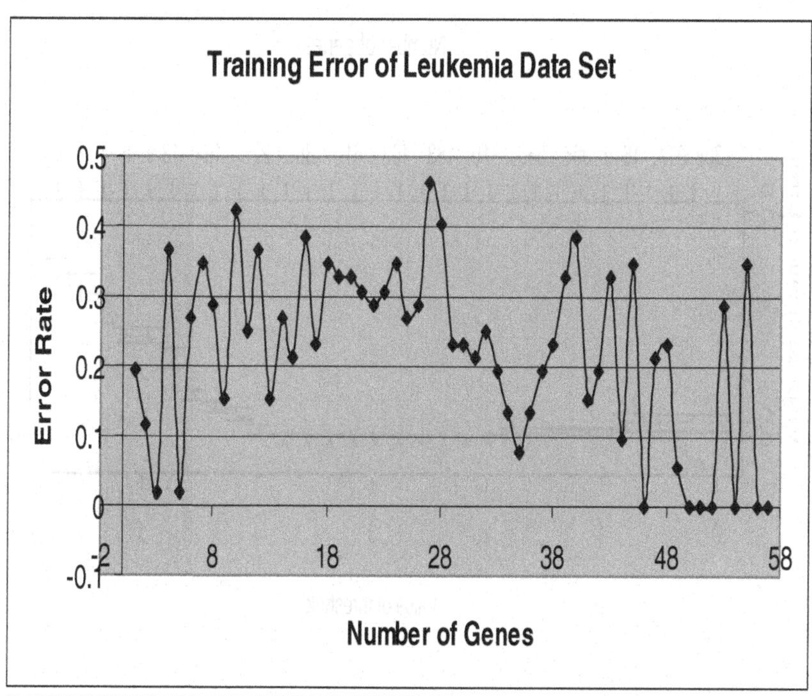

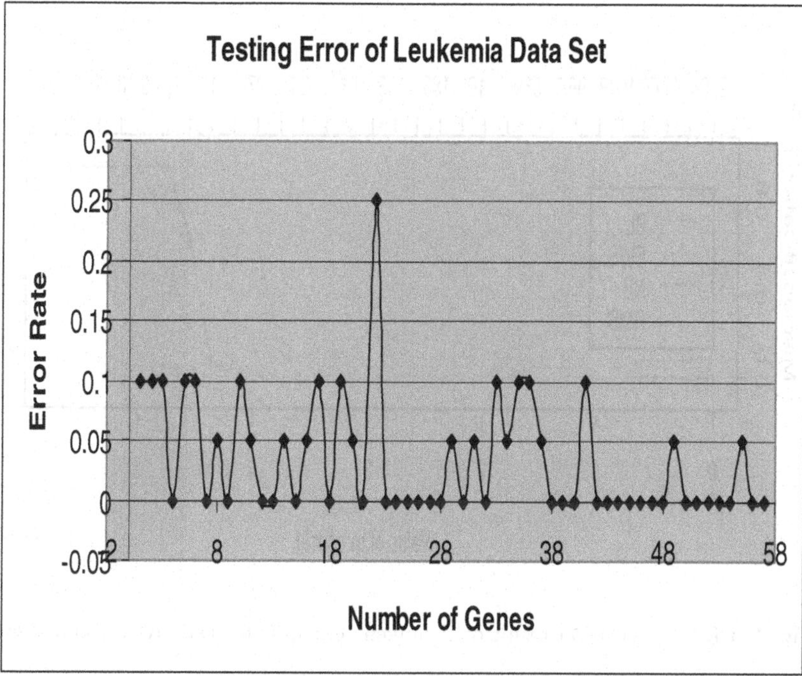

Fig. 3.15. the FNN classification result for the leukemia data set, with 100% training and testing accuracy using only 46 genes.

Number of genes

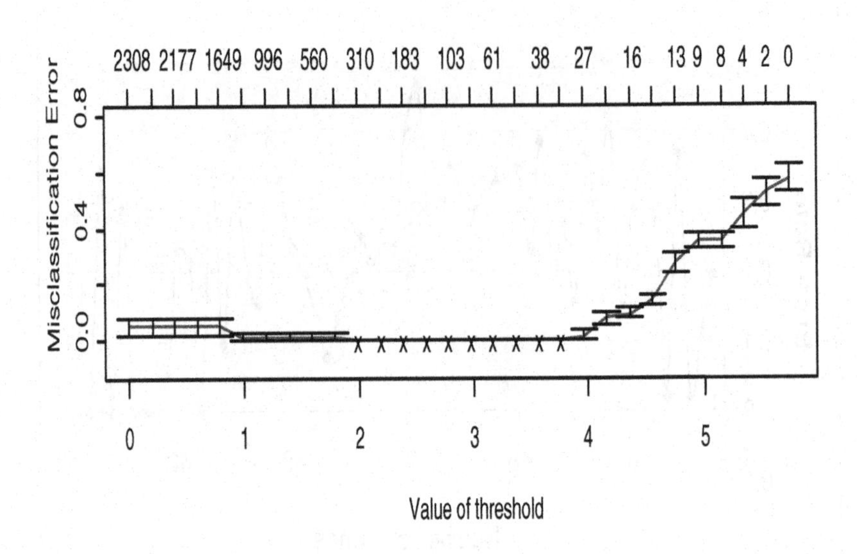

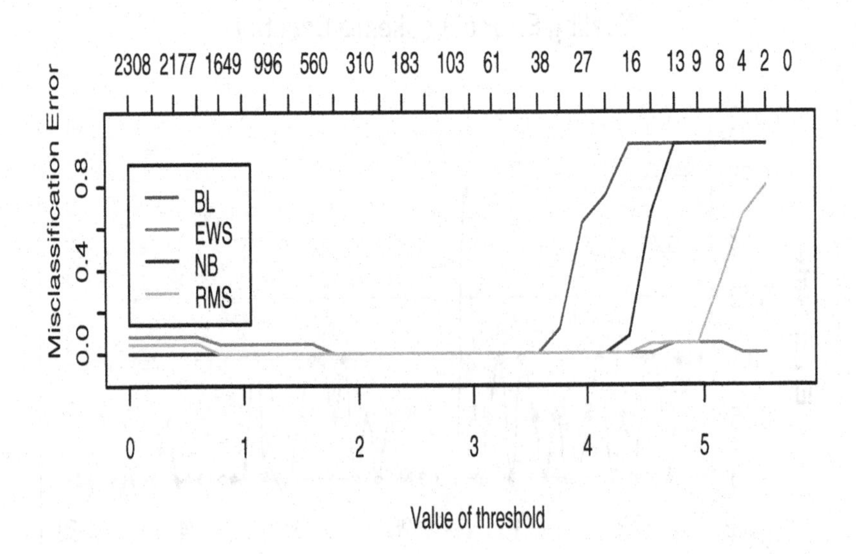

Fig. 3.16. Classification result of shrunken centroids for the SRBCT data set.

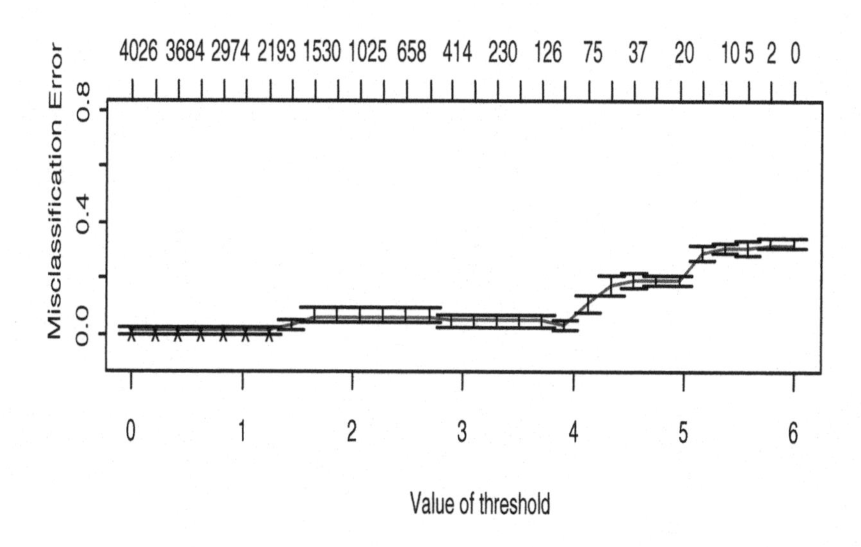

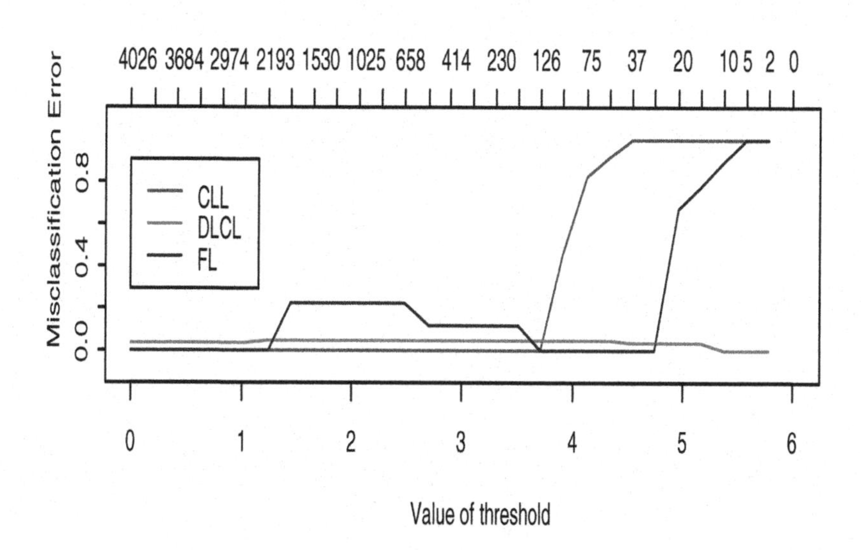

Fig. 3.17. Classification result of shrunken centroids for the lymphoma data set.

4

An Improved RBF Neural Network Classifier

4.1 Introduction

RBF neural networks [36][91][167][189][217][219][224][240] have been applied to channel equalization, image classification, function approximation, interpolation, density estimation, and classification tasks.

The popularity of the RBF neural network is due to its fast training and its global approximation capability with local responses. As with other neural networks, constructing an RBF neural network with high classification rates and a compact architecture but a high generalization capability is a challenging task. It is desirable that a computational model trained on a set of samples has both high learning and generalization capabilities. Learning ability measures how well a learning model approximates the functional relationship between the input and the output samples used in training, whereas generalization capability refers to how well the learning model processes new samples, not used during training. The training error rate E_{tr} is the error rate that a trained model makes for the training data, whereas the test error rate E_{te} is the error rate that a learning model makes for the test data. E_{tr} and E_{te} are used to measure the learning and the generalization capabilities of a learning model, respectively.

The number of hidden units is an important parameter for constructing RBF neural networks. Maffezzoni and Gubian [203] set the initial number of hidden neurons as the number of training samples and then used a manual pruning procedure to obtain an optimal number of hidden neurons. In [17][36], trials were carried out with several prespecified numbers of hidden neurons and the best number of hidden neurons was then selected based on the trial results. However, it is difficult to prespecify the number of hidden units or prune the RBF architecture from a large scale. In Roy *et al.* [264]'s algorithm, the number of hidden neurons was determined dynamically while training the RBF neural network, but the number of hidden units is still large. There is room to reduce the complexity of RBF neural networks by reducing the number of hidden units. Two useful modifications [107] to the algorithm of Roy *et al.* [264] are as follows:

1. Large overlaps are allowed between clusters of the same class, which reduce the number of hidden units without degrading the performance of RBF neural networks.
2. Dynamic determination of the overlaps between different classes.

Our simulations show that these modifications bring about improvements in terms of both structural simplicity and classification accuracy.

4.2 RBF Neural Networks for Classification

In a classification task, for each class C_i $(i = 1, 2, ..., M)$ with N_i samples, a posterior probability is [22]:

$$P(C_i|X) = \frac{p(X|C_i)P(C_i)}{p(X)} = \frac{p(X|C_i)P(C_i)}{\sum_{i'=1}^{N} p(X|C_{i'})P(C_{i'})}, \tag{4.1}$$

where X is a data sample. $\sum N_i = N$.

The probability mentioned above can be represented as an RBF neural network with radial basis kernel functions given by [22]:

$$\o_i(X) = \frac{p(X|C_i)}{\sum_{i'=1}^{N} p(X|C_{i'})P(C_{i'})} \tag{4.2}$$

and second-layer connections which consist of one weight from each hidden unit to the corresponding output unit, with value $p(C_i)$. Thus, the posterior probabilities can be approximated by the RBF neural network.

An RBF neural network may be considered as a mixture model for representing the distribution of the data set. Assume that K radial basis kernel functions are generated in an RBF neural network in order to represent all the class-conditional densities. The jth kernel function can be expressed as follows [22]:

$$p(X|C_i) = \sum_{j=1}^{K} p(X|j)P(j|C_i). \tag{4.3}$$

The unconditional density is:

$$p(X) = \sum_{i=1}^{N} p(X|C_i)P(C_i)$$
$$= \sum_{j=1}^{K} p(X|j)P(j), \tag{4.4}$$

where the probabilities of the radial basis functions are:

$$P(j) = \sum_{i=1}^{N} P(j|C_i)P(C_i). \tag{4.5}$$

The unconditional density of the input data ($p(X)$) can be presented by a mixture model, in which the component densities are given by the radial basis functions shown in Eq. (4.4).

We obtain the posterior probabilities of class membership by substituting Eq. (4.3) and Eq. (4.4) into Eq. (4.1) [22]:

$$p(C_i|X) = \frac{\sum_{j=1}^{K} P(j|C_i)p(X|j)P(C_i)}{\sum_{j'=1}^{K} p(X|j')P(j')} \frac{P(j)}{P(j)}$$
$$= \sum_{j=1}^{K} w_{ij}\o_j(X). \tag{4.6}$$

Equation (4.6) represents an RBF neural network with K radial basis kernel functions, in which the weights are given by [22]:

$$w_{ij} = \frac{P(j|C_i)P(C_i)}{P(j)}$$
$$= P(C_i|j) \tag{4.7}$$

and the jth kernel function is given by [22]:

$$\o_j(X) = \frac{P(X|j)P(j)}{\sum_{j'=1}^{K} p(X|j')P(j')}$$
$$= P(j|X). \tag{4.8}$$

Thus, the classification of an RBF neural network can be interpreted as the posterior probabilities represented by the radial basis functions, and the weights connecting the hidden units with the outputs.

4.2.1 The Pseudo-inverse Method

For a neural network classifier, training is based on its classification performance. The MSE (mean squared error) function is usually used as the objective function in neural networks:

$$E = \frac{1}{2}|d - Y|^2, \tag{4.9}$$

where d is the target vector and Y is the output vector.

In an RBF neural network, its MSE function is as follows:

$$E = \frac{1}{2}\sum_{n=1}^{N}\sum_{m=1}^{M}\{\sum_{j=0}^{K} w_{mj}\o_j^n - t_m^n\}^2, \tag{4.10}$$

where N is the number of patterns, M is the number of outputs, and K is the number of hidden units. w_{mj} is the weight connecting the jth hidden unit with the mth output unit. $\o_j^n$ represents the output of the jth kernel function for the nth input pattern. t_m^n represents the target output of the mth output unit when inputting the nth pattern.

Assume that the parameters (the number of hidden units, centers, and widths of hidden units) of the hidden layer have been fixed at the first training stage. weights between the hidden layer and the output layer need to be determined. In order to minimize the MSE, Eq. (4.16) is differentiated with respect to w_{mj} and the derivative is set to be zero [22]:

$$\sum_{n=1}^{N}\{\sum_{j'=0}^{K} w_{mj'}\o_{j'}^n - t_m^n\}\o_j^n = 0. \tag{4.11}$$

Equation (4.11) is written in the form of a matrix:

$$(\phi^T\phi)W^{\mathrm{T}} = \phi^{\mathrm{T}}T, \tag{4.12}$$

where ϕ, with elements $\o_j^n$, has dimensions $N \times K$. W is an $M \times K$ matrix with elements w_{mj}. T has dimensions $N \times M$ and elements t_m^n. The matrix $\phi^T\phi$ in Eq. (4.12) is a square matrix with dimensions $K \times K$. If $\phi^{\mathrm{T}}\phi$ is a non-singular matrix, the solution to Eq. (4.12) is given [120][253] as follows:

$$W^{\mathrm{T}} = \phi^{\dagger}T, \tag{4.13}$$

where $\phi^{\dagger}$ is a $K \times N$ matrix known as the pseudo-inverse of ϕ:

$$\phi^{\dagger} \equiv (\phi^{\mathrm{T}}\phi)^{-1}\phi^{\mathrm{T}}. \tag{4.14}$$

It is noted that, if the matrix $\phi^{\mathrm{T}}\phi$ is singular, Eqs. (4.13) and (4.14) do not exist, i.e., there is not a unique solution for Eq. (4.12). Redefine the pseudo-inverse of ϕ as:

$$\phi^{\dagger} \equiv lim_{\epsilon \to 0}(\phi^{\mathrm{T}}\phi + \epsilon I)^{-1}\phi^{\mathrm{T}}, \tag{4.15}$$

where I is the unit matrix. $\epsilon > 0$. It is clear that the limit always exists [22], and it can be used to minimize the MSE.

4.2.2 Comparison between the RBF and the MLP

Though both the RBF neural network [33][85][219][245] and the MLP neural network are powerful tools in function approximation, classification, and data mining tasks, there exist differences in their performances and applications, which result from the differences stated as follows:

1. Activation function [22].
   In RBF neural networks, the activation of a hidden unit is determined by the transformed distance between the input pattern and the center of the hidden unit. The transformation of the distance is by an activation function with a localized nature, such as the Gaussian kernel function. The activations of hidden units of MLP neural networks depend on weighted linear summations of the inputs transformed by activation functions, such as the sigmoidal function and the hyperbolic tangent function, which are not local in nature.
2. Data space partition.
   An MLP neural network separates the data space by hyper-planes, while an RBF neural network generates hyper-spheres to partition the input space. This difference is a direct consequence of the difference in activation functions used in the RBF and MLP networks.
3. Training procedure.
   The parameters of an MLP neural network are usually determined in a single training procedure. An RBF neural network's parameters are typically trained in several stages. The parameters of the kernel functions are trained first by unsupervised techniques, and the weights connecting the hidden layer and the output layer are determined at the second stage.
4. The weights.
   All weights in an MLP neural network are usually adjustable. The weights between the input layer and the hidden layer in a typical RBF neural network are fixed as 1's. The weights connecting hidden units and output units can be obtained by the linear least square (LLS) method.
5. The local minima problem.
   The error function to be minimized may have numerous local minima in the parameter space. The weights between the hidden layer and the output layer in an RBF network can be determined by the LLS method, which does not lead to any local minima; however, hidden neuron parameter adjustments by clustering may cause local minima.

4.3 Training a Modified RBF Neural Network

Our main objective is to discover hidden information from data sets, and represent the discovered information in an explicit and understandable way. A small RBF neural network classifier with high generalization capability is desirable in implementing data mining tasks.

Finding the centers, widths, and weights connecting hidden nodes with the output is the key to constructing and training the RBF classifier.

Overlapped receptive fields of hidden neurons for different classes can improve the performance of the RBF classifier when dealing with noisy data [203]. Kaylani *et al.* [170] and Bishop *et al.* [264] created overlapping Gaussian kernel functions (clusters) to map out the territory of each class with a smaller number of Gaussian functions.

In those previous methods, the clusters are formed as follows. A pattern is randomly selected from the data set V as the initial center of a cluster. The radius of this cluster is chosen in such a way that the ratio between the number of patterns of a certain class (in-class patterns) and the total number of patterns in the cluster is not less than a predefined value θ. Once this cluster is formed, all patterns inside this cluster are 'removed' from the training data set and do not participate in the formation of other clusters. The value of θ is determined empirically and is related to an acceptable classification error rate. Since θ determines the radii of the clusters, it also indirectly determines the degree of overlaps between different classes. Generally, a large θ leads to small radii of clusters; thus, it leads to small overlaps between the Gaussians for different classes and a small classification error rate for the training data set. Since a small classification error is desired, there usually exist small overlaps between the Gaussians representing the clusters.

Let us consider a simple example. Suppose that $\theta = 0.8$, i.e., there must be at least 80% in-class patterns in each cluster. In Fig. 4.1(a), suppose that cluster A has been formed and its members 'removed' from the data set V. Suppose that pattern 2 is subsequently selected as the initial center of a new cluster and cluster B is thus formed. Clusters C through G are then formed subsequently in a similar fashion. We see that clusters B, C, and D are quite small and therefore the effectiveness of the above clustering algorithm needs to be improved.

An algorithm [107] is used to reduce the number of clusters as follows. We first make a copy V_c of the original data set V. When a qualified cluster (with the ratio of in-class patterns higher than θ), e.g., cluster A in Fig. 4.1(b) (same as in Fig. 4.1(a)), is generated, the members in this cluster are 'removed' from the copy data set V_c, but the patterns in the original data set V remain unchanged. Subsequently, the initial center of the next cluster is selected from the *copy data set* V_c, but the candidate members of this cluster are patterns in the *original data set* V, and thus include the patterns in the cluster A. When pattern 2 is selected as an initial cluster center, a much larger cluster B, which combines clusters B, C, and D in Fig. 4.1(a), can still meet the θ-criterion and can therefore be created. By allowing for large overlaps between clusters for *the same class*, we can further reduce the number of clusters substantially. This will lead to more efficient construction of RBF networks, and will be demonstrated by computer simulations in the next section.

Overlaps between clusters of *different* classes affect the classification error rate, i.e., the larger the overlaps between clusters of *different* classes, often the larger the classification error rate. However, when large overlaps between clusters of the *same* class are allowed, it is not expected to degrade the classification accuracy. Since large overlaps between clusters of the same class can help combine small clusters into larger ones, and noise may thus be suppressed by these combinations, the accuracy of classification may even be improved compared with the results without the modification.

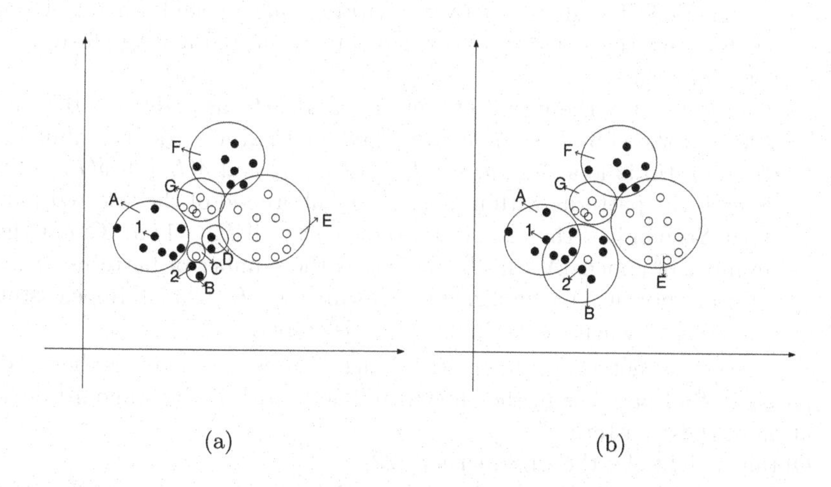

(a) (b)

Fig. 4.1. A comparison between (a) existing algorithms: small overlaps between clusters and (b) the modified algorithm with a reduced number of clusters: small overlaps between clusters of different classes, but large overlaps between clusters of the same class. (© 2005 IEEE) We thank the IEEE for allowing the reproduction of this figure, first appeared in [100].

We therefore use the following algorithm to construct an efficient RBF classifier, incorporating the above modification to the existing algorithms [264][265]:

1. Initialize for training:
   a) We divide the data set into three parts, the training data set, the validation data set, and the test data set.
   b) In order to derive the widths of the kernel functions, a general scale of neighborhood δ_0 is obtained by calculating the standard deviation of the data set [264].

2. Set phase $L = 1$ (L indicates the steps in one training epoch); $\delta(L) = \delta_0$, $\delta(1) = \alpha\delta_0$, where $\delta(L)$ is the initial radius of clusters at the training phase L (the training phase indicates the status of the training epoch, in which the initial radii of the clusters are affected), and δ is the increment step for the radius. α is the change rate of the radius.
3. Generate V_c, a copy of the original training data set V.
4. Forming clusters:
   a) Count the number of patterns in classes in V_c. If the number of patterns in a class is fewer than a predefined number (which is determined empirically), the patterns in the class will not be selected.
   b) Set subphase L_s=1 (L_s indicates the shrinking degree of the radius of cluster in a subphase training procedure), $\delta(L_s) = \delta(L)$.
   c) Select randomly a pattern from V_c as an initial cluster center and search in V for all the patterns within a $\delta(L)$-neighborhood of the center pattern. Thus, large overlaps are permitted among clusters of the same class.
   d) Check whether the ratio between in-class patterns and the total patterns in the subset is equal to or greater than a predefined value θ. If the ratio is less than θ, set $L_s = L_s + 1$, and $\delta(L_s) = \delta(L_s) - \delta$. Search the patterns within a $\delta(L_s)$-neighborhood of the selected pattern. Stop only if the ratio criterion is met or if $L_s \geq 1/2\alpha$. Count the number of epochs N_e, if $N_e \geq Di$ (i is the number of patterns in V_c, D is an integer. D is empirically determined.), $\theta = 0.95\theta$. Repeat step 4, and stop when the training set V_c is empty.
5. Calculate the center and the width of each cluster: the center is the mean pattern of all patterns in the cluster and the width is the standard deviation of these patterns.
6. Obtain weights by the LLS method [22].
7. Calculate E_{tr} (the classification error of the training set) and E_v (the classification error of the validation set). Stop if both E_{tr} and E_v are smaller than a prespecified value E_0. E_0=E_{Pre} (E_{Pre} is a predefined value of classification error rate. E_{Pre}=2%). Else:
   If $E_v(L) < E_v(L-1)$, set $L = L+1$ and $\delta(L) = \delta(L) - \delta/2$. Go to step 3.
   If $E_{tr}(L) > E_{tr}(L-1)$ and $E_v(L) > E_v(L-1)$, $L = L+1$, set $\delta(L) = \delta(L) - \delta$. Go to step 3.
   If $L > 1/\alpha$ or $E_v > E_0$ Go to step 2.

Compared to Roy *et al.*'s original algorithm [264][265], we have made the following changes:

1. In step 4, large overlaps among clusters of the same class are allowed in order to reduce the number of hidden units without reducing the classification accuracy.

2. In [264], six training phases were used corresponding to θ={50%, 60%, ..., 100%}, respectively. In each phase, θ is unchanged and all clusters of this phase should meet the θ-criterion. A classification result is obtained for this θ. The classification error rate from a phase is compared to that in the previous phase. Whether to continue to the next training phase is determined by a comparison in the classification results. If the classification accuracy is better than the previous phase, training continues with a higher θ, i.e., it is possible to obtain a higher accuracy with θ increased. Thus, the data is trained continuously using a larger θ. Otherwise, the training is stopped. Although a larger θ leads to a higher accuracy in the training data set, it may lead to poorer generalization in the test data set. In our algorithm (step 2), θ is automatically adjusted according to the training condition, i.e., how many patterns of the class concerned are left. With the decreasing number of patterns, θ is decreased by a certain factor, say 0.95 in our simulations.

The aim of the modification stated above, i.e., allowing for large overlaps among clusters of the same class, is to decrease the number of Gaussian hidden units without reducing the classification accuracy.

The cost of the modification is increased training time. Assume that the number of patterns in the data set is N. The number of hidden units is M_1 when allowing for large overlaps among clusters of the same class, and the number of hidden units is M_2 without allowing for large overlaps. For a certain hidden unit k, the number of patterns in this cluster is N_k. The processing time for one pattern is T in the training procedure. Assume that both algorithms (with the modification and without the modification) have the same initial conditions. The initial center of each candidate cluster is selected randomly. Thus, the average number of trials in the two algorithms for searching for qualified clusters may be assumed to be the same, say M_0. With the modification, all N patterns in the data set will be checked when searching for a qualified cluster in each trial. The time required for one trial is thus NT. For M_0 trials, the total time required is M_0NT. Without the modification, if a cluster is considered to be qualified, the patterns included in this cluster will be removed from the data set. Thus, the total time required for classification without the modification is $\sum_{m=1}^{M_0} T(N - \sum_{k=1}^{m-1} N_k) = (M_0N - \sum_{m=1}^{M_0} \sum_{k=1}^{m} N_k)T < M_0NT$. Thus, more time is needed when applying the modification. However, the cost is worthwhile in that less complicated classifiers with higher accuracy are found in most cases.

4.4 Experimental Results

The Iris, Monk3, Thyroid, Mushroom, and Breast cancer data sets from the UCI Repository of Machine Learning Databases [223] are used here to demonstrate the above algorithms. Each data set is divided into three parts, i.e., training, validation, and test sets. We set $\alpha = 0.1$ and initial $\theta = 100\%$ in our experiments. Each experiment is repeated five times with different initial conditions and the average results are recorded. In each run of five implementations, a data set is randomly divided into three subsets. Around 60% data points are used for training, 20% for testing, and 20% for validation.

4.4.1 Iris Data Set

The Iris data set records the physical dimensions of three kinds of Iris flowers. There are four attributes and 150 patterns in the Iris data set. In the 150 patterns, 90 patterns are for training, 30 patterns for validation, and 30 patterns for testing.

The four attributes are sepal length, sepal width, petal length and petal width, which are all in centimeters. There are three classes, i.e., Iris Setosa, Iris Versicolour, and Iris Virginica.

Table 4.1 shows that when large overlaps among clusters of the same class are permitted, the number of hidden units is reduced from 4.6 to 3.4 on average, and the classification error rate is maintained.

Table 4.1. Reduction in the number of hidden neurons and classification error rates of the RBF neural network for the Iris data set when large overlaps among clusters with the same class label are permitted (average results of five independent runs).

Performance criteria	Iris	
	Small overlaps	Large overlaps
Training error	0%	0%
Validation error	0%	0%
Testing error	3.33%	3.33%
Number of hidden neurons	4.6	3.4

4.4.2 Thyroid Data Set

The Thyroid data set records the medical test results of patients with thyroid diseases. There are five attributes and 215 patterns in the Thyroid data set, with 115 data points in the training set, 50 data points in the testing set, and 50 data points in the validation set.

The five attributes are T3-resin uptake test (a percentage), total serum thyroxin as measured by the isotopic displacement method, total serum triiodothyronine as measured by radioimmuno assay, basal thyroid-stimulating hormone (TSH) as measured by radioimmuno assay, and the maximal absolute difference of the TSH value after injection of 200 micrograms of thyrotropin-releasing hormone as compared to the basal value. The three classes are normal, hyper-thyroid, and hypo-thyroid.

Table 4.2 shows that when large overlaps among clusters of the same class are permitted, the number of hidden units is reduced from 14.4 to 8 on average, and the classification error rate is reduced from 6.0% to 4.8%.

Table 4.2. Same as Table 4.1, with the Thyroid data set.

Performance criteria	Thyroid	
	Small overlaps	Large overlaps
Training error	3.65%	2.78%
Validation error	4.8%	6%
Testing error	6.0%	4.8%
Number of hidden neurons	14.4	8

4.4.3 Monk3 Data Set

The Monk3 data set is a collection of a binary classification problem over a six-attribute discrete domain. The data set involves learning a binary function defined over this domain, from some training examples of this function.

There are 122 patterns in the training set and 432 patterns in the test set. We divide the test set into 200 patterns for validation and 221 patterns for testing since the size of the validation set affects the generalization performance of RBF classifiers.

Table 4.3 shows that when large overlaps among clusters of the same class are permitted, the number of hidden units is reduced from 33.2 to 19.6 on average, and the classification error rate increases slightly from 5.91% to 6.88%.

Table 4.3. Same as Table 4.1, with Monk3 data set.

Performance criteria	Monk3	
	Small overlaps	Large overlaps
Training error	4.10%	4.92%
Validation error	8.0%	5.76%
Testing error	5.91%	6.88%
Number of hidden neurons	33.2	19.6

4.4.4 Breast Cancer Data Set

This breast cancer data set was obtained from Dr. William H. Wolberg of the University of Wisconsin Hospitals, Madison. There are nine attributes and 699 patterns in the Breast cancer data set. 16 patterns with missing attributes are removed. Of the 683 patterns left, 444 were benign and the rest were malignant. In the 683 patterns, 410 patterns are for training, 136 for validation, and 137 for testing.

The attributes are clump thickness, uniformity of cell size, uniformity of cell shape, marginal adhesion, single epithelial cell size, bare nuclei, bland chromatin, normal nucleoli, and mitoses. There are two classes, i.e., benign and malignant.

Table 4.4 shows that when large overlaps among clusters of the same class are permitted, the number of hidden units is decreased from 31 to 11 on average, and the classification error rate is reduced from 2.92% to 1.46%.

Table 4.4. Same as Table 4.1, with the breast cancer data set.

Performance criteria	Breast Cancer	
	Small overlaps	Large overlaps
Training error	2.44%	0.73%
Validation error	2.92%	1.96%
Testing error	2.92%	1.46%
Number of hidden neurons	31	11

4.4.5 Mushroom Data Set

The mushroom data set includes descriptions of hypothetical samples corresponding to 23 species of gilled mushrooms in the Agaricus and Lepiota families.

There are 22 nominal attributes and 8124 patterns in the Mushroom data set. Among the 8124 patterns, 4500 patterns are for training, 1812 are for validation, and 1812 are for testing. There are two classes, edible and poisonous.

The attributes are:

1. cap-shape: bell (b), conical (c), convex (x), flat (f), knobbed (k), sunken (s)
2. cap-surface: fibrous (f), grooves (g), scaly (y), smooth (s)
3. cap-color: brown (n), buff (b), cinnamon (c), gray (g), green (r), pink (p), purple (u), red (e), white (w), yellow (y)

4. bruises?: bruises (t), no (f)
5. odor: almond (a), anise (l), creosote (c), fishy (y), foul (f), musty (m), none (n), pungent (p), spicy (s)
6. gill-attachment: attached (a), descending (d), free (f), notched (n)
7. gill-spacing: close (c), crowded (w), distant (d)
8. gill-size: broad (b), narrow (n)
9. gill-color: black (k), brown (n), buff (b), chocolate (h), gray (g), green (r), orange (o), pink (p), purple (u), red (e), white (w), yellow (y)
10. stalk-shape: enlarging (e), tapering (t)
11. stalk-root: bulbous (b), club (c), cup (u), equal (e), rhizomorphs (z), rooted (r), missing (?)
12. stalk-surface-above-ring: fibrous (f), scaly (y), silky (k), smooth (s)
13. stalk-surface-below-ring: fibrous (f), scaly (y), silky (k), smooth (s)
14. stalk-color-above-ring: brown (n), buff (b), cinnamon (c), gray (g), orange (o), pink (p), red (e), white (w), yellow (y)
15. stalk-color-below-ring: brown (n), buff (b), cinnamon (c), gray (g), orange (o), pink (p), red (e), white (w), yellow (y)
16. veil-type: partial (p), universal (u)
17. veil-color: brown (n), orange (o), white (w), yellow (y)
18. ring-number: none (n), one (o), two (t)
19. ring-type: cobwebby (c), evanescent (e), flaring (f), large (l), none (n), pendant (p), sheathing (s), zone (z)
20. spore-print-color: black (k), brown (n), buff (b), chocolate (h), green (r), orange (o), purple (u), white (w), yellow (y)
21. population: abundant (a), clustered (c), numerous (n), scattered (s), several (v), solitary (y)
22. habitat: grasses (g), leaves (l), meadows (m), paths (p), urban (u), waste (w), woods (d)

Table 4.5 shows that when large overlaps among clusters of the same class are permitted, the number of hidden units is decreased from 35 to 29 on average, and the classification error rate remains.

Table 4.5. Same as Table 4.1, with the Mushroom data set.

Performance criteria	Mushroom	
	Small overlaps	Large overlaps
Training error	0.8%	1.0%
Validation error	0.5%	0.7%
Testing error	1.1%	1.1%
Number of hidden neurons	35	29

4.5 RBF Neural Networks Dealing with Unbalanced Data

4.5.1 Introduction

In neural network training, if some classes have much fewer samples compared with the other classes, the neural network system may respond wrongly for the minority classes because the overwhelming samples in the majority classes dominate the adjustment procedure in training.

In [19], a cost-sensitive neural network was proposed, in which different costs were associated with making errors in different classes. When the sum of squared errors was calculated for the multi-player perceptron (MLP) neural network, each term was multiplied by a class-dependent factor (cost). This idea has a much earlier origin in machine learning, that is, the loss matrix [22] which deals with different risks (costs) associated with making errors in different classes. For example, in classification of medical images, these class-dependent risk factors will need to be selected from practical experiences. By assigning larger costs to minority classes, Berardi and Zhang [19] were able to improve the classification accuracies for minority classes. In that work, as in earlier discussions on risk matrices, cost factors are selected in an ad hoc manner. In this work, we propose a method that determines these cost factors automatically such that all classes, i.e., minority and majority classes, are roughly equally important. We demonstrate this method in the case of the RBF classifier.

In [201], two methods had been presented for handling unbalanced data sets. In the first method, the samples of minority classes were duplicated to increase their effects on training neural networks. In the second method, the so-called snowball method proposed in [330] for multi-font character recognition was used to improve the accuracy of the minority class [201]. In the snowball method, neural networks were first trained with the samples in the minority class, which favors the minority populations. Next, samples of majority classes were added gradually while training the neural network dynamically. The two methods were used in the MLP neural network, the ART (adaptive resonance theory) neural network, and the RBF neural network. However, it was found that the two methods mentioned above had no effect on the MLP and RBF neural networks.

We explore a training method of RBF neural networks for handling unbalanced data [103]. Constructing an RBF neural network with a compact architecture but robust for unbalanced data is a challenging task.

In this section, we present an algorithm, first proposed in [103], which deals with unbalanced data by increasing the contribution of minority samples to the MSE (mean squared error) function. The MSE function is a powerful objective function in neural network training. When learning a data set with a minority class, the weights of a neural network will be dominated by the majority classes. For example, when using the back-propagation method to train weights, the weights are updated according to the variation in the error function of the neural network [201]. It is clear that the weights obtained finally will reflect the nature of the majority classes but not much of the minority class. Thus, it motivates us to increase the magnitudes of weighted parameters of minority classes to balance the influence of the unbalanced classes, i.e., the errors brought by different classes are weighted by parameters inversely proportional to the number of samples in the classes.

4.5.2 The Standard RBF Neural Network Training Algorithm for Unbalanced Data Sets

For a neural network classifier, its training algorithm is developed based on its classification performance. The MSE function (Eq. (4.9)) is usually used as the objective function in neural networks.

Here the MSE function of an RBF neural network is written as follows:

$$E_0(W) = \frac{1}{2}\sum_{n=1}^{N}\sum_{m=1}^{M}\{y_m^n - t_m^n\}^2. \qquad (4.16)$$

where W is the weight vector, N is the number of patterns, M is the number of outputs, and K is the number of hidden units. w_{mj} is the weight connecting the jth hidden unit with the mth output unit. $\o_j^n$ represents the output of the jth kernel function for the nth input pattern. t_m^n represents the target output of the mth output unit when inputting the nth pattern.

Assume that there are M classes in a data set and M output neurons in the network. The mth output of an RBF neural network corresponding to the nth input vector is as follows:

$$y_m^n(\mathbf{X}^n) = \sum_{j=1}^{K} w_{mj}\o_j(\mathbf{X}^n) + w_{m0}b_m. \qquad (4.17)$$

Here, $\mathbf{X}^n$ is the nth input pattern vector, $m = 1, 2, ..., M$, and K is the number of hidden units. w_{mj} is the weight connecting the jth hidden unit to the mth output node. b_m is the bias. w_{m0} is the weight connecting the bias and the mth output node. $\o_j^n(\mathbf{X}^n)$ is the activation function of the jth hidden unit corresponding to the nth input vector.

$$\o_j^n(\mathbf{X}^n) = \mathrm{e}^{\frac{-||\mathbf{X}^n - \mathbf{C_j}||^2}{2\sigma_j{}^2}}, \qquad (4.18)$$

where $\mathbf{C}_j$ and σ_j are the center and the width for the jth hidden unit respectively, which are adjusted during learning. When calculating the distance between input patterns and centers of hidden units, Euclidean distance measure is employed in RBF neural networks.

Substitute Eq. (4.17) into Eq. (4.16):

$$E_0(W) = \frac{1}{2}\sum_{n=1}^{N}\sum_{m=1}^{M}\{(\sum_{j=0}^{K} w_{mj}\o_j^n + w_{m0}b_m) - t_m^n\}^2. \qquad (4.19)$$

Differentiate E_0 with respect to w_{mj} and let

$$\frac{\partial E_0(W)}{\partial w_{mj}} = 0, \qquad (4.20)$$

then as shown in earlier section, Eq. (4.11) is obtained. Equation (4.11) is written in a matrix notation (Eq. (4.12)), which leads to the *pseudo-inverse* for solving weights.

4.5.3 Training RBF Neural Networks on Unbalanced Data Sets

It is shown in the above equations that there is no particular attention paid to unbalanced cases, in which the sample sizes of different classes in a data set are unbalanced. Unbalanced training data may lead to an unbalanced architecture in training. In our work, we add larger weights on minority classes in order to attract more attention in training for the minority members.

Assume that the number of samples in class i is N_i. The total number of samples in the data set is $N = N_1 + \cdots + N_i + \cdots + N_M$. The error function shown in Eq. (4.19) can be written as:

$$E_0(W) = \frac{1}{2}\sum_{i=1}^{M}\sum_{n_i=1}^{N_i}\sum_{m=1}^{M}\{(\sum_{j=0}^{K} w_{mj}\o_j^{n_i} + w_{m0}b_i) - t_m^{n_i}\}^2. \qquad (4.21)$$

During the training of neural networks with unbalanced training data, a general error function such as Eq. (4.16) or Eq. (4.21) cannot lead to a balanced classification performance on all classes in the data set because majority classes contribute more compared to minority classes and therefore result in more weight adjustments on majority classes. In supervised training algorithms, neural networks are constructed by minimizing a neural network error function whose variables are the network weights connecting layers. Thus, the training procedure has a bias towards frequently occurring classes.

In order to increase the contribution of minority classes in weight adjustments, we change Eq. (4.21) to:

$$E(W) = \frac{1}{2}\sum_{i=1}^{M}\beta_i\sum_{n_i=1}^{N_i}\sum_{m=1}^{M}\{(\sum_{j=0}^{K} w_{mj}\o_j^{n_i} + w_{m0}b_i) - t_m^{n_i}\}^2, \qquad (4.22)$$

where

$$\beta_i = \frac{N}{N_i}, i = 1, 2, ..., M. \qquad (4.23)$$

Differentiate E with respect to w_{mj}, and let

$$\frac{\partial E(W)}{\partial w_{mj}} = 0. \tag{4.24}$$

Substituting Eq. (4.22) into Eq. (4.24), we obtain:

$$\sum_{i=1}^{M} \beta_i \sum_{n_i=1}^{N_i} \{\sum_{j'=0}^{K} w_{mj'} \o_{j'}^{n_i} - t_m^{n_i}\} \o_j^{n_i} = 0. \tag{4.25}$$

We introduce a new parameter r_n replacing β_i:

$$r_n = \beta_i \quad \text{when} \quad \mathbf{X}^n \in A_i. \tag{4.26}$$

A_i is class i. Substitute Eq. (4.26) into Eq. (4.25):

$$\sum_{n=1}^{N} r_n \{\sum_{j'=0}^{K} w_{mj'} \o_{j'}^{n} - t_m^{n}\} \o_j^{n} = 0. \tag{4.27}$$

By replacing r_n with $\sqrt{r_n}\sqrt{r_n}$, we obtain:

$$\sum_{n=1}^{N} \{\sum_{j'=0}^{K} w_{mj'} \o_{j'}^{n}.\sqrt{r_n} - t_m^{n}\sqrt{r_n}\} \o_j^{n}\sqrt{r_n} = 0. \tag{4.28}$$

Similarly as stated in [22], there is the following new pseudo-inverse equation for calculating weight W:

$$(\phi^{\mathrm{T}}\phi)W^{\mathrm{T}}T = \phi^{\mathrm{T}}T. \tag{4.29}$$

Different to the pseudo-inverse equation shown in Eq. (4.12), here $\phi \rightarrow \o_j^n\sqrt{r_n}$, and $T \rightarrow t_i^n\sqrt{r_n}$.

As indicated in the above equations, we have taken the unbalanced data into consideration when training RBF neural networks. The parameters r_n introduce biased weights which are opposite to the proportions of classes in a data set. The effect of the weight parameters r_n is shown in Sect. 4.5.4. Compared with the training method without considering an unbalanced condition in the data, the classification accuracy of the minority classes is improved. We also allow large overlaps between clusters of the same class to reduce the number of hidden units [102][104].

The modified training algorithm for RBF neural networks, in which small overlaps between clusters of different classes and large overlaps between clusters of the same class are allowed, is used in this section.

4.5.4 Experimental Results

The car evaluation data set in Chap. 3 are used here to demonstrate our algorithm. The data set is divided into three parts, i.e., training, validation, and test sets. Each experiment is repeated five times with different initial conditions and the average results are recorded.

We generate an unbalanced car data set based on function 5 shown in Chap. 3. There are nine attributes and two classes: Class A and Class B. Samples which do not meet the conditions of Class A are samples of Class B in the car data set. 4000 patterns are in the training data set and 2000 patterns for the testing data set. There are 507 patterns of class 1 (Class A) in the training data set, and 205 patterns of class 1 in the testing data set. The testing data set is divided into two subsets: the validation set and the testing set with 1000 patterns, respectively. Class A is the minority class. Class B is the majority class.

Comparison between small overlaps and large overlaps among clusters of the same class are shown on classification error rates and the number of hidden units. When allowing large overlaps among clusters of the same class, the number of hidden units is reduced from 328 to 303, and the classification error rate on the test data set is increased slightly from 4.1% to 4.5%.

In table 4.6, the comparison of overall classification error rates between with and without considering the unbalanced condition is shown. Here large overlaps are allowed between clusters with the same class label. It is also shown in Table 4.6, when considering the unbalanced condition in the data set, that the classification error rate of the minority class decreases from 34.65% to 8.73%. At the same time, the error rate of the majority class increases slightly from 1.37% to 4.1%. Since, in most cases, the minority class is embedded with important information, improving the individual accuracy of the minority class is critical.

In this section, a modification [103] is described to the training algorithm for the construction and training of the RBF network on unbalanced data by increasing bias towards the minority classes. Weights inversely proportional to the number of patterns of classes are given to each class in the MSE function. Experimental results show that the proposed method is effective in improving the classification accuracy of minority classes while maintaining the overall classification performance.

4.6 Summary

In this chapter, we described a modified training algorithm for RBF neural networks, which we proposed earlier [107]. This modified algorithm leads to fewer hidden units while maintaining the classification accuracy of RBF classifiers. Training is carried out without knowing in advance the number of hidden units and without making any assumptions on the data.

We described two useful modifications to Roy *et al.*'s algorithm for the construction and training of an RBF network, by allowing for large overlaps among clusters of the same class and dynamically determining the cluster overlaps of different classes.

Table 4.6. Comparison of classification error rates of the RBF neural network for each class of the car data set between with and without considering the unbalanced condition when allowing large overlaps between clusters with the same class label (average results of five independent runs).

Without considering unbalanced condition		
Overall error rates		
Training set	Validation set	Testing set
1.89%	5.0%	4.8%
Class A		
Training set	Validation set	Testing set
11.69%	27.69%	34.65%
Class B		
Training set	Validation set	Testing set
0.77%	2.41%	1.37%
Considering unbalanced condition		
Overall error rates		
Training set	Validation set	Testing set
1.2%	5.1%	4.5%
Class A		
Training set	Validation set	Testing set
4.27%	4.58%	8.73%
Class B		
Training set	Validation set	Testing set
0.85%	5.15%	4.1%

In RBF neural network classifiers, larger overlaps between different classes lead to higher classification errors. However, large overlaps between clusters with the same class labels will not degrade classification performance since the overlaps occur between clusters of the same class, i.e., the number of hidden units, is reduced and the classification error rate is reduced or maintained by this modification.

The ratio between the number of patterns of a certain class (in-class patterns) and the total number of patterns in the cluster represents the overlaps of different classes. A dynamic parameter θ is applied to control the ratio according to the training condition. If the trials for searching for a qualified cluster reach a certain threshold, θ will be decreased for searching clusters.

The two modifications may help reduce detrimental effects from noisy patterns and isolated patterns while maintaining classification performance. There are two training stages in the training algorithm. By searching for clusters based on the proposed modifications, widths, and centers of Gaussian kernel functions are determined at the first training stage. Weights connecting the hidden layer and the output layer are determined at the second training stage by the LLS method. Experimental results show that the modifications are effective in reducing the number of hidden units while maintaining or even increasing the classification accuracy.

This new approach can be feasibly used for classification when the underlying distributions of the data are unknown. The accuracy is comparable with Roy *et al.*'s method [264], but the computational time is greater than Roy's method. However, based on the experimental results, there seems to be room for further research to speed up the training algorithm. In future work, the present approach could be enhanced by analyzing the relationships among the clusters for improving classification accuracies and reducing computational time.

In addition, a new algorithm is presented for the construction and training of an RBF neural network with unbalanced data. In applications, minority classes with much fewer samples are often present in data sets. The learning process of a neural network is usually biased towards classes with majority populations. Our study focused on improving the classification accuracy of minority classes while maintaining the overall classification performance.

5

Attribute Importance Ranking for Data Dimensionality Reduction

Large-scale data can only be handled with the aids of computers. However, processing commands may need to be entered manually by data analysts and data mining results can be fully used by decision makers only when the results can be understood explicitly. The removal of irrelevant or redundant attributes could benefit us in making decisions and analyzing data efficiently. Data miners are expected to present discovered knowledge in an easily understandable way. Data dimensionality reduction (DDR) is an essential part in the data mining processes. Drawn from methods in pattern recognition and statistics, DDR is developed to fulfill objectives such as improving accuracy of prediction models, scaling the data mining models, reducing computational cost, and providing a better understanding of knowledge extracted.

5.1 Introduction

DDR plays an important role in data mining tasks since those semi-automated or automated methods perform better with lower-dimensional data with the removal of irrelevant or redundant attributes compared to higher-dimensional data. Irrelevant or redundant attributes as unuseful information often interfere with useful ones. In the classification task, the main aim of DDR is to reduce the number of attributes used in classification while maintaining an acceptable classification accuracy.

The problem of DDR is to select a subset of attributes which represents the concept of data without losing important information. Feature (attribute) extraction and feature selection are two techniques of DDR. LDA (linear discriminant analysis) [168][198] and PCA (principal component analysis) [166] are common feature extraction methods. However, by the transformation operation in feature extraction, new features which are linear or non-linear combinations of the original features are generated. Unwanted artifacts often come out with the new features. In addition, non-linear transformation is usually not reversible, which brings difficulties in understanding data through extracted features.

Feature selection does not generate unwanted artifacts, i.e., feature selection is carried out in the original measurement space. This can be achieved by removing redundant or irrelevant attributes without losing the original concept of data.

In optimal feature selection, all possible feature combinations should be inspected. Though some methods are explored to save some work [43], the high computational cost is still a problem unsolved. Under the circumstance, suboptimal feature selection algorithms are an alternative. Though suboptimal feature selection algorithms do not guarantee the optimal solution, the selected feature subset usually leads to a higher performance in the induction system (such as a classifier).

One wishes to find a measure that can determine the irrelevant attributes with little computational cost. Consider two samples with different class labels in a data set, which are presented by a set of attributes. There are differences observed in the two samples' attributes, i.e., there are correlations between attributes and class labels. Irrelevant attributes will not reflect the correlation relationship when changing from one sample to another sample, and the correlations may be used to rank attribute importance.

On the other hand, large class distance is expected in order to distinguish different classes. Irrelevant attributes have no positive influence on separating distinct classes, and the removal of redundant attributes has no negative influence on forming distinct classes. Hence, class separability can be used as a criterion to evaluate attribute importance.

Feature selection can be performed based on the evaluation of attribute importance. Dash *et al.* [71] proposed an entropy measure to rank attribute importance. In mutual information based feature selection (MIFS) [18][27], 'the information content' of each attribute (feature) is evaluated corresponding to classes and the other attributes. However, the number of attributes included in the selected attribute subset has to be predefined, which requires prior knowledge of data. The importance level of attributes is evaluated by the evaluation criterion. Kononenko [180] introduced a Relief-F method to rank attribute importance in order to reduce data dimensionality. In the Relief-F method, for a given instance, nearest neighbors are searched for from each class. The difference in an attribute in each pair of instances is calculated. The importance level of the attribute is evaluated by the probabilities of these differences.

In this chapter, we describe a novel separability-correlation measure (SCM), which was first proposed in [107] for determining the importance of the original attributes. Then, different attribute subsets obtained based on attribute ranking results are used as inputs to RBF classifiers. The classification results are used to evaluate the feature subsets in order to reduce the data dimensionality, and the RBF network architecture can be simplified with the reduced attribute subsets. The SCM includes two parts, the intraclass distance to interclass distance ratio and an attribute-class correlation measure. The attribute-class correlation measure is used to evaluate the power of each attribute affecting the class label for each pattern. The larger the correlation factor, the more important the attribute is for determining the class labels of patterns. The ratio of the intraclass distance and the interclass distance reflects the class separability. The relative importance of a feature is given by its relative magnitude of the SCM.

5.2 A Class-Separability Measure

The farther apart the classes are, the easier it is to classify them. Therefore, to identify a subset of features that can maximize the separability between classes is a desirable objective of feature selection.

For example, two classes C_1 and C_2 are shown in Fig. 5.1. The average pairwise distance between patterns of the two classes reflects the separability of the two classes, i.e., the greater the average pairwise distance, the better the separability of the two classes [76]. When the number of patterns is large, the cost of the pairwise-distance calculation is high. We used the average interclass distance $S_b = \sum_{i=1}^{C} P_i[(\mathbf{m}_i - \mathbf{m})(\mathbf{m}_i - \mathbf{m})^{\mathrm{T}}]^{\frac{1}{2}}$ to replace the pairwise distance (see Fig. 5.2) [76]. This can be easily introduced to multiple-class data. Here C is the number of classes in the data set and P_i is the probability of the ith class. $\overline{\mathbf{X}}_{ik}$ is the normalized data vector, whose jth attribute, $\overline{X}_{ik}(j)$ is normalized as:

$$\overline{X}_{ik}(j) = \frac{X_{ik}(j) - \min(x_j)}{\max(x_j) - \min(x_j)}, \tag{5.1}$$

where $\max(x_j)$ and $\min(x_j)$ are the maximum and the minimum of the jth attribute in the data set, respectively. Equation (5.1) can normalize data to the range [0, 1]. $j = 1, 2, ..., n$. n is the number of attributes. $X_{ik}(j)$ is the original (un-normalized) data. $\mathbf{m}_i$ is the mean vector of the ith class:

$$\mathbf{m}_i = \frac{\sum\limits_{k=1}^{n_i} \overline{\mathbf{X}}_{ik}}{n_i}. \tag{5.2}$$

$\mathbf{m}$ is the mean of all patterns in the data set:

$$\mathbf{m} = \frac{\sum\limits_{i=1}^{c} \sum\limits_{k=1}^{n_i} \overline{\mathbf{X}}_{ik}}{n}, \tag{5.3}$$

where n_i is the number of patterns in the ith class. N is the total number of patterns in the data set, i.e., $N = n_1 + n_2 + \cdots + n_c$.

The intraclass distance S_w represents distances of patterns within a class. $S_w = \sum_{i=1}^{C} P_i/n_i \sum_{k=1}^{n_i}[(\overline{\mathbf{X}}_{ik} - \mathbf{m}_i)(\overline{\mathbf{X}}_{ik} - \mathbf{m}_i)^T]^{\frac{1}{2}}$ may reflect the density of data within a class. The denser the data within each class, the easier it is to

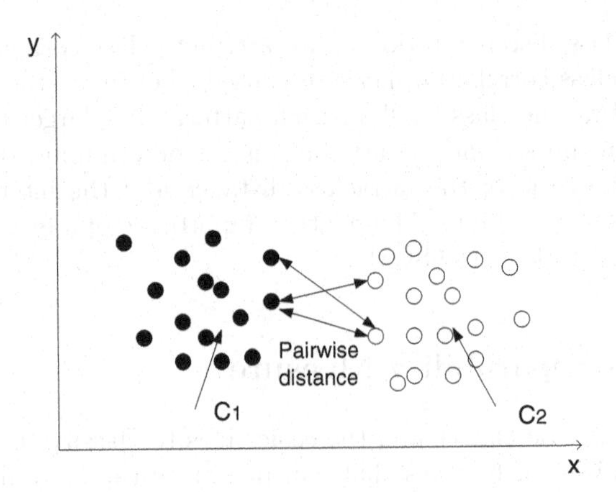

Fig. 5.1. Pairwise interclass relationship.

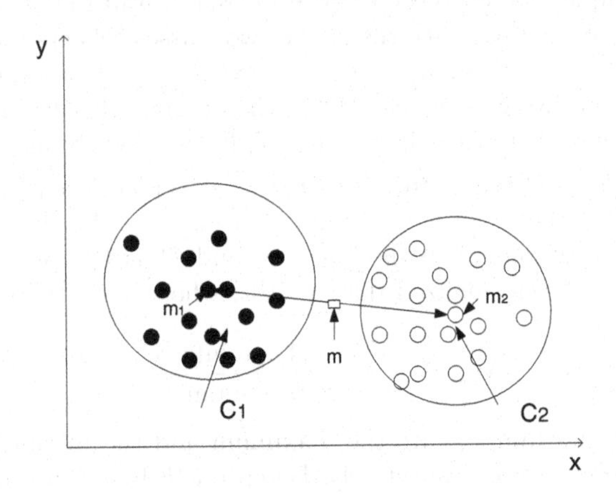

Fig. 5.2. Average interclass relationship.

classify. Thus, class-separability may be measured by the intraclass distance S_w and the interclass distance (the distance between patterns of different classes) S_b.

The greater S_b is and the smaller S_w is, the better the separability of the data set is. Therefore, the ratio of S_w and S_b can be used to measure the distinction of the classes: the smaller the ratio, the better the separability.

If removing attribute k_1 from the data set leads to less class separability, i.e., a greater S_w/S_b, compared to the case where attribute k_2 is removed, one may consider attribute k_1 more important for classification of the data set than attribute k_2 is, and vice versa. Hence, we may rank the importance of the attributes by calculating the intraclass-to-interclass distance ratio with each attribute omitted in turn.

However, the ratio S_w/S_b does not always work well as a class-separability measure. For example, consider two classes, with one class surrounding the other, but which are completely *separable*. Since $\mathbf{m}_1$, $\mathbf{m}_2$, and $\mathbf{m}$ defined in Eq. (5.2) and Eq. (5.3) are equal, $S_b \to 0$, which indicates total *inseparability*. Here there is a need to have other importance measures.

5.3 An Attribute-Class Correlation Measure

In addition to the separability of classes in the data set, the correlation between the changes in attributes and their corresponding changes in class labels should be taken into account when ranking the importance of attributes. This correlation directly links features with class labels.

If the class labels of two patterns are different, the variations of attributes in the two patterns are considered to be the affecting factor for the variations of class labels and should be weighted positively. If the class labels of two patterns are the same, the variations in the attributes are irrelevant in deciding the classes and should be weighted negatively. The correlation measure can be a useful factor when combined with the class-separability measure.

We describe the following correlation [107] between the kth attribute and the class labels in the data set:

$$C_k = \sum_{i \neq j} |\overline{\mathbf{X}}_{ik} - \overline{\mathbf{X}}_{jk}| \mathrm{magn}(y_i - y_j), \tag{5.4}$$

where $\overline{\mathbf{X}}_{ik}$ and $\overline{\mathbf{X}}_{jk}$ are the kth attributes of the ith pattern and the jth pattern, respectively. y_i and y_j are the class labels of the ith pattern and the jth pattern, respectively. For any y, $\mathrm{magn}(y) = 1$ if $|y| > 0$ and $\mathrm{magn}(y) = -0.05$ if $|y| = 0$, which can help enlarge the differences of importance among attributes, i.e., if the magnitude of an attribute changes from one data pattern to another though the class label is unchanged, then a negative value should apply to the attribute for evaluating it. A great magnitude of C_k shows that there is a close correlation between class labels and the kth attribute, which indicates the great importance of attribute k in classifying the patterns, and vice versa.

5.4 The Separability-correlation Measure for Attribute Importance Ranking

The following separability-correlation measure (SCM) [107] that we proposed earlier is used to evaluate the importance levels of attributes by combining the above two measures:

$$R_k = \chi \overline{S_k} + (1 - \chi) \overline{C_k}, \tag{5.5}$$

where $S_k = S_{wk}/S_{bk}$, and $\overline{S_k} = (S_k - \min(S_k))/(\max(S_k) - \min(S_k))$ is the normalization of S_k. $\max(S_k)$ and $\min(S_k)$ are the maximum and minimum of all S_k, respectively. $k = 1, 2, ..., n$. n is the number of attributes. S_{wk} and S_{bk} are intraclass and interclass distances calculated with the kth attribute omitted from each pattern, respectively. For example, the ith pattern $\mathbf{X}_i = \{x_{i1}, x_{i2}, ...x_{ik}, x_{ik+1}, ..., x_{in}\}$ becomes $\mathbf{X}_i'=\{x_{i1}, x_{i2}, ..., x_{ik-1}, x_{ik+1}, ..., x_{in}\}$ when R_k is calculated. $\overline{C_k} = (C_k - \min(C_k))/(\max(C_k) - \min(C_k))$ is the normalization of C_k. χ is a weight parameter; $1 \geq \chi \geq 0$ and χ is determined empirically: the best choice of χ should lead to a subset of attributes which results in the highest classification accuracy.

The importance levels of attributes are ranked using the values of R_k. The greater the magnitude of R_k, the more important the kth attribute. We will demonstrate the use of our SCM method in Sect. 5.7.

We use a combination of two measures, i.e., class separability and attribute-class correlation, because either of them alone does not work well, as shown in our experimental results presented later in the book.

5.5 Different Searches for Ranking Attributes

Bottom-up, Top-down, and Exhaustive searches can be used for ranking attributes. In a bottom-up search, we begin from an empty set. The SCM is used for evaluating each attribute by omitting this attribute from each pattern, i.e., the attribute is considered to be more important if its corresponding SCM magnitude is larger than others. The selected attribute is included in the empty attribute subset. This operation is continued until n attributes are included. The order of attributes entering the attribute set indicates the importance order of attributes. In a bottom-up search, the number of attribute combinations in a SCM calculation for determining attribute importance equals n (n is the number of attributes).

It starts from the complete attribute set in a top-down search. Each attribute is removed from the attribute set temporarily for calculating its SCM. Then the least important attribute, whose corresponding value in SCM is the smallest, is eliminated from the current attribute set. The steps are iterated until only one attribute is left in the attribute set. The number of attribute combinations in the SCM calculation is $n + (n-1) + \cdots + 1 = n(n+1)/2$. In Sect. 5.7, the differences of the two searches will be shown.

Thus, more attribute combinations are considered in the top-down search than in the bottom-up search, which may lead to different feature selection results since some combinations may not be inspected in the bottom-up search.

All the possible attribute subsets are examined in an exhaustive search. The number of attribute combinations needed to be checked is $C_n^1 + C_n^2 + \cdots C_n^{n-1} + C_n^n = 2^n - 1$. In the branch and bound algorithm, some attribute sets need not be examined, which leads to a saving in computation; however, the number of calculations still grows exponentially with n. In addition, the computational saving is achieved by assuming that the feature selection evaluation function is monotonic [76].

Due to the computational burden of optimal search methods, one has to resort to suboptimal feature selection methods. In classification tasks, since the goal is to obtain better classification accuracy with less complicated construction of classifiers, the strategy of using the classification accuracy as evaluation for selecting features is used widely. Suboptimal search is used and RBF classifiers are used as evaluators.

5.6 Data Dimensionality Reduction

Based on whether or not the feature selection is carried out independently of induction algorithms, DDR can be categorized into the filter approach [187] and the wrapper approach [158][246]. The weakness of the filter approach lies in that the selected feature subset may not lead to high performance in induction systems, such as the classification system. And, the wrapper approach combines DDR with induction algorithms, but high computational cost is a heavy burden.

Classification of patterns may be based on a very few of the most important attributes. Determining which attributes should be retained for the original concept of data is pivotal in feature selection techniques. The basic idea to select a subset of attributes from measure space is to inspect all the possible subsets. The best subset is selected on the basis that its value is greatest for a criterion function which represents the concept of data. This idea is theoretically satisfactory, but the computation burden is heavy in practice. It motivates people to rank the importance level of each attribute first and then choose the attribute subset according to the order of attributes. For example, for a six-dimensional data set, in order to obtain a best subset of attributes, in the basic feature selection method one needs to calculate $2^6 = 64$ times. But by ranking the importance level first, we only need six calculations to determine the selected subset.

In feature selection, we try to avoid selecting too many or too few features than necessary. If insufficient features are selected, the information content to keep the concept of the data is degraded. If too many features are selected, including redundant or irrelevant features, the classification accuracies may be lower due to the interference of irrelevant information. In data mining applications, the removal of irrelevant features is significant for discovering hidden relationships between features, and between features and class targets of patterns.

Unsuitable reduction of data dimensionality results in loss of information and may degrade the quality of data mining tasks, such as the classification task. In classification, the evaluator of DDR is usually the classifier. The RBF neural network classifier can be used as an evaluator for selecting features. Our algorithm is based on the fact that the removal of irrelevant or redundant attributes from the attribute set may not change the original concept of data, but the removal of important attributes will change the original concept of data.

Assume that the data space has L dimensions. The aim of the DDR task is to select the best subset X' with l ($l < L$) dimensions ($X' = \{x'_i \mid i = 1, 2, ..., l, x'_i \in X\}$) from the original attribute set X ($X = \{x_i \mid i = 1, 2, ..., L\}$). The best feature subset is composed of l features which optimize a criterion function, $A(\cdot)$, i.e., the selected feature subset X' satisfies [76]:

$$A(X') = \max_{Y} A(Y), \tag{5.6}$$

where Y refers to any other feature combination, $Y = \{y_i \mid i = 1, 2, ..., G, G = 1, 2, ..., L-1, y_i \in X\}$. L is the original dimensionality of data.

The criterion function $A(\cdot)$ usually has with the following characteristics [76]:

1. $A(\cdot) \geq 0$
2. $A(\cdot)$ has its maximum value when the data classes in X' space are disjoint.
3. For feature $x_i \in X$, there are the following inequalities:

$$A(x'_1) \geq A(x'_2) \geq \cdots \geq A(x'_l) \geq \cdots \geq A(x'_L), \tag{5.7}$$

where the original feature set $X = \{x'_i \mid \forall i \leq L\}$.

In a classification task, if there are infinite samples provided to classifiers, the detrimental effect of irrelevant or redundant features on the classification performance can be ignored. However, in practical applications, the number of samples is finite. Under this circumstance, the error brought in cannot be negligible. DDR is usually used to facilitate the classification task and improve the performance of classification.

5.6.1 Simplifying the RBF Classifier Through Data Dimensionality Reduction

Both the data dimensionality and the distribution of the input patterns affect the number of hidden units in RBF neural networks. If the data dimensionality is reduced, the number of hidden units will also be decreased in many cases. We use an RBF classifier together with the SCM to select the best subsets of attributes.

Based on the attribute importance ranking, we describe how to reduce the structural complexity and improve the performance of the RBF network as follows. According to the rank of attribute importance level obtained by the algorithm described in this chapter, J most important attributes are used as inputs of the RBF neural network classifier for $J = 1, 2, ..., N-1, N$. The classification error rateis calculated for each J. Thus, N classification error rates are calculated corresponding to N subsets of selected attributes. For small J, the classification error rate decreases as J increases until all relevant attributes are included. As J increases further, the classification error rate may remain unchanged or even increase because redundant or irrelevant attributes are included. The best subset of attributes is the one with the smallest classification error rate.

5.7 Experimental Results

5.7.1 Attribute Ranking Results

The Iris, Monk3, Thyroid, Breast cancer, Ionosphere, and Mushroom data sets from the UCI Repository of Machine Learning Databases [223] are used to demonstrate our algorithms for ranking attribute importance and constructing a simplified RBF network.

The Ionosphere data set is radar data which were collected in Goose Bay, Labrador. This system consists of a phased array of 16 high-frequency antennas with a total transmitted power on the order of 6.4 KW. There are 34 attributes and two classes in the Ionosphere data set.

For Iris, Monk3, Thyroid, and Breast cancer data sets, attribute importance rankings using the SCM with different χ's (Eq. (5.5)) are shown in Table 5.1, which shows that χ affects the order of attribute importance ranking because different χ's change the weights of the class-separability measure and the attribute-class correlation measure. Five χ's are used, i.e., $\chi = 0.0$, 0.4, 0.5, 0.7, and 1.0. In order to determine which order is better, different subsets of attributes are inputs to the RBF classifier for each order, so as to find the best subset for that order. If there are n original attributes in the data set, there are n candidate subsets of attributes as discussed in Sect. 5.5. The classification results are used to evaluate the attribute subsets. We select the subset of attributes corresponding to the lowest classification error rate for each data set and each ranking order.

According to the experimental results, when $\chi = 0.4$, the importance ranking results for the first four data sets lead to the lowest or nearly the lowest validation error rates with the smallest attribute subsets.

For the first four data sets, the comparison between importance-ranking results obtained by our SCM using bottom-up and top-down searches when $\chi = 0.4$, SUD [71], and Relief-F [180] is shown in Table 5.28.

For the Mushroom data set, attribute ranking queues with different χ's are shown in Table 5.17. $\chi = 0$ corresponds to the lowest classification error rate in the validation set with the smallest attribute set for the Mushroom data set. The comparison between importance ranking results obtained by our SCM using bottom-up and top-down searches when $\chi = 0$, SUD, and Relief-F is shown in Table 5.18 for the Mushroom data set.

For the Ionosphere data set, attribute ranking queues are shown in Table 5.21. $\chi = 0.5$ corresponds to the lowest classification error rate in the validation data set with the smallest attribute set for Ionosphere.

Table 5.1. Attribute importance ranking using the SCM with different χ's obtained by bottom-up search. (© 2005 IEEE) We thank the IEEE for allowing the reproduction of this table, first appeared in [107].

χ	Iris	Monk3	Thyroid	Breast
$\chi = 0.0$	4,3,1,2	5,4,2,1,6,3	2,3,5,1,4	7,2,4,3,8,9,5,6,1
$\chi = 0.4$	4,3,1,2	5,2,4,1,6,3	2,3,5,4,1	2,7,3,4,9,5,8,6,1
$\chi = 0.5$	4,1,3,2	5,2,4,1,6,3	2,3,5,4,1	2,7,3,4,9,5,1,8,6
$\chi = 0.7$	1,4,2,3	5,2,4,1,6,3	2,5,3,4,1	2,7,1,3,4,9,5,8,6
$\chi = 1.0$	1,2,4,3	5,2,3,6,4,1	2,5,3,4,1	1,2,7,3,4,9,5,8,6

In the following subsections, the classification results are shown for each data set with different feature subsets as inputs of classifiers based on attribute ranking results.

5.7.2 Iris Data Set

In Tables 5.2 to 5.5, classification error rates are shown for all attribute subsets corresponding to different attribute importance ranking results based on different χ's. $\chi = 0.4$ is selected since it leads to the smallest attribute subset $\{3, 4\}$ with the nearly lowest classification error rate.

Table 5.2. Classification error rates for Iris data set with different subsets of attributes according to the importance ranking shown in Table 5.1 when $\chi = 0.0$ and $\chi = 0.4$. The attribute subset with the lowest validation error is highlighted in bold. (© 2005 IEEE) We thank the IEEE for allowing the reproduction of this table, first appeared in [107].

Attributes used	Error rate		
	Training set	Validation set	Test set
4	0.1222	0.0667	0.1333
4,3	**0.0333**	**0.0000**	**0.0333**
4,3,1	0.0556	0.0333	0.1000
4,3,1,2	0.0889	0.1000	0.1000

Table 5.3. Same as Table 5.2, when $\chi = 0.5$. (© 2005 IEEE) We thank the IEEE for allowing the reproduction of this table, first appeared in [107].

Attributes used	Error rate		
	Training set	Validation set	Test set
1	0.3333	0.2333	0.4333
1,4	0.0778	0.0000	0.1000
1,4,2	**0.0556**	**0**	**0.0333**
1,4,2,3	0.0556	0	0.0333

Table 5.4. Same as Table 5.2, when $\chi = 0.7$. (© 2005 IEEE) We thank the IEEE for allowing the reproduction of this table, first appeared in [107].

Attributes used	Error rate		
	Training set	Validation set	Test set
4	0.3333	0.3667	0.4667
4,1	0.1111	0.2000	0.1000
4,1,3	0.3333	0.2333	0.4333
4,1,3,2	**0.0778**	**0.1000**	**0.0333**

Table 5.5. Same as Table 5.2, when $\chi = 1.0$. (© 2005 IEEE) We thank the IEEE for allowing the reproduction of this table, first appeared in [107].

Attributes used	Error rate		
	Training set	Validation set	Test set
1	0.4556	0.5667	0.3667
1,2	0.1889	0.3333	0.2333
1,2,4	0.0778	0.0667	0.1667
1,2,4,3	**0.0444**	**0.0667**	**0.0333**

5.7.3 Monk3 Data Set

In Tables 5.6 to 5.8, classification error rates are shown for all attribute subsets corresponding to different attribute importance ranking results based on different χ's. $\chi = 0.4$ is selected since it leads to the smallest attribute subset $\{2, 4, 5\}$ with the lowest classification error rates.

5.7.4 Thyroid Data Set

In Tables 5.9 and 5.10, classification error rates for all attribute subsets corresponding to different attribute importance ranking results based on different

Table 5.6. Classification error rates for Monk3 data set with different subsets of attributes according to the importance ranking shown in Table 5.1 when $\chi = 0.0$. The attribute subset with the lowest validation error is highlighted in bold. (© 2005 IEEE) We thank the IEEE for allowing the reproduction of this table, first appeared in [107].

Attributes used	Error rate		
	Training set	Validation set	Test set
5	0.2705	0.2328	0.2100
5,4	0.2541	0.3060	0.2450
5,4,2	**0.0902**	**0.0991**	**0.0650**
5,4,2,1	0.1967	0.2371	0.2050
5,4,2,1,6	0.1148	0.0948	0.1000
5,4,2,1,6,3	0.1885	0.2112	0.2600

Table 5.7. Same as Table 5.6, when $\chi = 0.4$, $\chi = 0.5$, and $\chi = 0.7$. (© 2005 IEEE) We thank the IEEE for allowing the reproduction of this table, first appeared in [107].

Attributes used	Error rate		
	Training set	Validation set	Test set
5	0.1880	0.3000	0.2870
5,2	0.1780	0.2830	0.2690
5,2,4	**0.0242**	**0.0585**	**0.067**
5,2,4,1	0.0899	0.3360	0.1830
5,2,4,1,6	0.0498	0.1897	0.1320
5,2,4,1,6,3	0.0328	0.2030	0.1240

χ's are shown. $\chi = 0.4$ is selected since it leads to the smallest attribute subset $\{2, 3, 5\}$ with the lowest classification error rates.

5.7.5 Breast Cancer Data Set

For 5 χ's, there are five different attribute importance ranking results. In Tables 5.11 to 5.15, classification error rates for all attribute subsets corresponding to different attribute importance ranking results based on different χ's are shown. $\chi = 0.4$ is selected since it leads to the smallest attribute subset $\{2, 3, 7\}$ with the lowest classification error rates.

5.7.6 Mushroom Data Set

Corresponding to five χ's, we obtain five different attribute importance ranking results for the Mushroom data set. In Table 5.19, the experimental results

Table 5.8. Same as Table 5.6, when $\chi = 1.0$. (© 2005 IEEE) We thank the IEEE for allowing the reproduction of this table, first appeared in [107].

Attributes used	Error rate		
	Training set	Validation set	Test set
5	0.2705	0.2328	0.2100
5,2	0.2213	0.1853	0.2050
5,2,3	0.1967	0.1638	0.1400
5,2,3,6	**0.1066**	**0.0690**	**0.0700**
5,2,3,6,4	0.2131	0.1767	0.1700
5,2,3,6,4,1	0.1230	0.1552	0.1600

Table 5.9. Classification error rates for the Thyroid data set with different subsets of attributes according to the importance ranking shown in Table 5.1 when $\chi = 0.0$. The attribute subset with the lowest validation error is highlighted in bold. (© 2005 IEEE) We thank the IEEE for allowing the reproduction of this table, first appeared in [107].

Attributes used	Error rate		
	Training set	Validation set	Test set
2	0.1860	0.1628	0.2093
2,3	0.0698	0.0698	0.2093
2,3,5	**0.0543**	**0.0465**	**0.0930**
2,3,5,1	0.0543	0.0465	0.1163
2,3,5,1,4	0.0388	0.0465	0.1395

Table 5.10. Same as Table 5.9, when $\chi = 0.4$, $\chi = 0.5$, $\chi = 0.7$, and $\chi = 1.0$. (© 2005 IEEE) We thank the IEEE for allowing the reproduction of this table, first appeared in [107].

Attributes used	Error rate		
	Training set	Validation set	Test set
2	0.0930	0.0930	0.0930
2,3	0.0698	0.0465	0.0698
2,3,5	**0.0543**	**0.0233**	**0.0233**
2,3,5,4	0.0543	0.0233	0.0465
2,3,5,4,1	0.0388	0.0233	0.0233

with and without removal of irrelevant attributes are compared for the Mushroom data set. $\chi = 0.0$ is selected since it leads to the smallest attribute subset $\{9, 20, 5\}$ with the lowest classification error rates shown in Table 5.20.

Table 5.11. Classification error rates for the Breast cancer data set with different subsets of attributes according to the importance ranking shown in Table 5.1 when $\chi = 0.0$. The attribute subset with the lowest validation error is highlighted in bold. (© 2005 IEEE) We thank the IEEE for allowing the reproduction of this table, first appeared in [107].

Attributes used	Error rate		
	Training set	Validation set	Test set
7	0.1100	0.0803	0.1022
7,2	0.0954	0.0803	0.0949
7,2,4	0.0391	0.0511	0.0219
7,2,4,3	**0.0318**	**0.0438**	**0.0073**
7,2,4,3,8	0.0367	0.0365	0.0219
7,2,4,3,8,9	0.0244	0.0365	0.0146
7,2,4,3,8,9,5	0.0318	0.0438	0.0146
7,2,4,3,8,9,5,6	0.0342	0.0365	0.0146
7,2,4,3,8,9,5,6,1	0.0342	0.0511	0.0219

Table 5.12. Same as Table 5.11, when $\chi = 0.4$. (© 2005 IEEE) We thank the IEEE for allowing the reproduction of this table, first appeared in [107].

Attributes used	Error rate		
	Training set	Validation set	Test set
2	0.1100	0.0803	0.1022
2,7	0.0709	0.0657	0.0876
2,7,3	**0.0269**	**0.0365**	**0.0073**
2,7,3,4	0.0391	0.0438	0.0365
2,7,3,4,9	0.0269	0.0365	0.0219
2,7,3,4,9,5	0.0342	0.0365	0.0146
2,7,3,4,9,5,8	0.0293	0.0438	0.0073
2,7,3,4,9,5,8,6	0.0269	0.0438	0.0146
2,7,3,4,9,5,8,6,1	0.0342	0.0365	0.0146

5.7.7 Ionosphere Data Set

There are 34 attributes for each data pattern of the Ionosphere data set. The total number of patterns is 351. 60% of the data points are used for training, 20% for validation, and 20% for testing. The comparison on the number of hidden units and classification error rates before and after removing irrelevant attributes is shown in Table 5.23. 21 attributes out of the original attribute set are selected corresponding to $\chi = 0.5$ as shown in Table 5.26. The error rate of the testing data set is decreased, and the number of hidden units is reduced as well. According to the ranking queue of attributes (seen in Table 5.22) obtained by the SUD method, the first 22 attributes in the attribute queue

Table 5.13. Same as Table 5.11, when $\chi = 0.5$. (© 2005 IEEE) We thank the IEEE for allowing the reproduction of this table, first appeared in [107].

Attributes used	Error rate		
	Training set	Validation set	Test set
2	0.1443	0.1314	0.1387
2,7	0.0513	0.0511	0.0365
2,7,3	**0.0269**	**0.0292**	**0.0073**
2,7,3,4	0.0318	0.0511	0.0146
2,7,3,4,9	0.0293	0.0438	0.0219
2,7,3,4,9,5	0.0269	0.0365	0.0292
2,7,3,4,9,5,1	0.0244	0.0438	0.0073
2,7,3,4,9,5,1,8	0.0318	0.0365	0.0146
2,7,3,4,9,5,1,8,6	0.0318	0.0365	0.0146

Table 5.14. Same as Table 5.11, when $\chi = 0.7$. (© 2005 IEEE) We thank the IEEE for allowing the reproduction of this table, first appeared in [107].

Attributes used	Error rate		
	Training set	Validation set	Test set
2	0.1443	0.1314	0.1387
2,7	0.0538	0.0584	0.0438
2,7,1	0.0685	0.0876	0.0730
2,7,1,3	**0.0342**	**0.0365**	**0.0146**
2,7,1,3,4	0.0367	0.0365	0.0219
2,7,1,3,4,9	0.0269	0.0438	0.0292
2,7,1,3,4,9,5	0.0318	0.0511	0.0146
2,7,1,3,4,9,5,8	0.0293	0.0365	0.0292
2,7,1,3,4,9,5,8,6	0.0391	0.0511	0.0292

are selected, which corresponds to the least classification error rate shown in Table 5.24. In Table 5.25, the results are shown corresponding to the attribute ranking queue of the Relief-F method. The feature subset with 21 attributes obtained based on our SCM leads to the classification error rate of 5.57%. The feature subset with 22 attributes obtained based on the SUD method leads to the classification error rate of 6.86%. And, the feature subset with 21 attributes obtained based on the Relief-F method leads to the classification error rate of 5.43%. The sizes of the feature subsets from the three attribute ranking methods are approximately equal to one another. The results derived from the Relief-F method correspond to a lower classification error rate.

Table 5.15. Same as Table 5.11, when $\chi = 1.0$. (© 2005 IEEE) We thank the IEEE for allowing the reproduction of this table, first appeared in [107].

Attributes used	Error rate		
	Training set	Validation set	Test set
1	0.3423	0.3869	0.3358
1,2	0.1467	0.1314	0.1460
1,2,7	0.0660	0.0730	0.0511
1,2,7,3	**0.0489**	**0.0292**	**0.0146**
1,2,7,3,4	0.0416	0.0365	0.0146
1,2,7,3,4,9	0.0367	0.0438	0.0292
1,2,7,3,4,9,5	0.0269	0.0365	0.0146
1,2,7,3,4,9,5,8	0.0318	0.0365	0.0219
1,2,7,3,4,9,5,8,6	0.0318	0.0365	0.0292

Table 5.16. Comparison of the numbers of hidden units and classification errors before and after irrelevant attributes are removed according to the SCM ranking method. B: before removal, A: after removal. (© 2005 IEEE) We thank the IEEE for allowing the reproduction of this table, first appeared in [107].

Comparison		Data set			
		Iris	Monk3	Thyroid	Breast
Input attributes	B	1,2,3,4	1,2,3,4,5,6	1,2,3,4,5	1,2,3,4,5,6,7,8,9
	A	4,3	5,2,4	2,3,5	2,7,3
Number of hidden units	B	4	19.6	8	11
	A	3	11.6	5	5
Classification error rate	B	0.0467	0.0688	0.0465	0.0146
	A	0.0333	0.067	0.0233	0.0073

Table 5.17. Attribute importance ranking for the Mushroom data set using the SCM with different χ's obtained by bottom-up search.

χ	Mushroom
$\chi = 0.0$	9, 20, 5, 3, 22, 1, 13, 14, 15, 11, 21, 12, 19, 2, 4, 8, 10, 7, 18, 17, 6, 16
$\chi = 0.4$	9, 20, 1, 5, 2, 13, 14, 11, 3, 22, 12, 15, 19, 21, 4, 8, 10, 7, 18, 17, 6, 16
$\chi = 0.5$	1, 20, 9, 2, 5, 13, 14, 11, 12, 19, 3, 22, 15, 21, 4, 8, 10, 7, 18, 17, 6, 16
$\chi = 0.7$	1, 20, 2, 9, 5, 13, 12, 19, 11, 14, 15, 22, 4, 3, 8, 21, 10, 7, 18, 17, 6, 16
$\chi = 1.0$	1, 2, 20, 13, 5, 9, 12, 19, 11, 4, 8, 14, 10, 7, 18, 6, 17, 16, 15, 22, 21, 3

5.7.8 Comparisons Between Top-down and Bottom-up Searches and with Other Methods

We compare results obtained from the SCM using bottom-up and top-down searches. We also compare with results derived from the attribute importance ranking by the Relief-F [180] and the SUD [71].

Table 5.18. Comparison for the Mushroom data set between importance ranking results obtained by our SCM using bottom-up and top-down searches when $\chi = 0.0$, the SUD, and the Relief-F methods.

Data set	Decreasing order of importance			
	SCM (bottom-up)	SCM (top-down)	SUD	Relief-F
Mushroom	9,20,5,3,22,1, 13,14,15,11,21, 12,19,2,4,8,10, 7,18,17,6,16	9,20,5,3,22,1, 13,14,15,11,21 12,19,2,4,8,10, 7,18,17,6,16	9,3,20,22,5,11, 21,2,19,15,14 1,13,12,4,10,8, 7,18,6,17,16	5,20,11,8,19,4, 10,22,9,12,13, 21,7,3,2,15,14, 18,6,17,16,1

Table 5.19. Comparison of the numbers of hidden units and classification errors before and after irrelevant attributes are removed according to the SCM ranking method for the Mushroom data set. B: before removal, A: after removal.

Comparison	Data set	
	Mushroom	
	B	A
Input attributes	1,2,3,4,5,6,7,8,9,10,11,12, 13,14,15,16,17,18,19,20,21,22	5,9,20
Number of hidden units	29	18
Classification error rate	1.1%	1.2%

Figure 5.3(a) shows the classification error rates of the RBF classifier for different subsets of Iris attributes according to the importance ranking obtained with the SCM using *bottom-up* search when $\chi = 0.4$. We obtained the same attribute ranking results and hence the same attribute subsets from the SCM using bottom-up and top-down searches (Table 5.28) for Iris. It is seen from Figs. 5.3(a) and (b) that as the number of attributes used increases, the test error first decreases, reaches a minimum when the first two attributes in the attribute ranking queue are used, and then increases. Hence, in the Iris data set, attributes 3 and 4 are relevant attributes for classification and are then selected, which improves the classification performance and decreases the number of inputs and the number of hidden units of the RBF neural network. The classification error rate is reduced from 0.0467 to 0.0333, and the number of Gaussian hidden units is reduced from 4 to 3 (Table 5.16 summarizes results for different data sets).

Further, we carry out classification by inputting different attribute sets to the RBF classifiers based on the attribute importance ranking results obtained by the SUD [71] and the Relief-F [180] methods (reproduced in Table 5.28). We compare the classification error rates of test data sets corresponding to the selected attribute subsets obtained using the SCM, SUD, and Relief-F methods (Table 5.27).

Table 5.20. Classification error rates for the Mushroom data set with different subsets of attributes according to the importance ranking shown in Table 5.1 when $\chi = 0.0$. The attribute subset with the lowest validation error is highlighted in bold.

Attributes used	Error rate		
	Training set	Validation set	Test set
9	0.027	0.044	0.030
9,20	0.035	0.047	0.022
9,20,5	**0.01**	**0.02**	**0.012**
9,20,5,3	0.083	0.042	0.060
9,20,5,3,22	0.041	0.081	0.035
9,20,5,3,22,1	0.092	0.080	0.071
9,20,5,3,22,1,13	0.079	0.082	0.078
9,20,5,3,22,1,13,14	0.068	0.097	0.068
9,20,5,3,22,1,13,14,15	0.0640	0.061	0.050
9,20,5,3,22,1,13,14,15,11	0.039	0.052	0.062
9,20,5,3,22,1,13,14,15,11,21	0.072	0.081	0.090
9,20,5,3,22,1,13,14,15,11,21,12	0.044	0.031	0.044
9,20,5,3,22,1,13,14,15,11,21,12,19	0.0490	0.053	0.034
9,20,5,3,22,1,13,14,15,11,21,12,19,2	0.039	0.035	0.045
9,20,5,3,22,1,13,14,15,11,21,12,19,2,4	0.025	0.036	0.024
9,20,5,3,22,1,13,14,15,11,21,12,19,2,4,8	0.023	0.046	0.020
9,20,5,3,22,1,13,14,15,11,21,12,19,2,4,8,10	0.033	0.025	0.024
9,20,5,3,22,1,13,14,15,11,21,12,19,2,4,8,10,7	0.023	0.038	0.032
9,20,5,3,22,1,13,14,15,11,21,12,19,2,4,8,10,7,18	0.022	0.029	0.048
9,20,5,3,22,1,13,14,15,11,21,12,19,2,4,8,10,7,18,17	0.039	0.039	0.035
9,20,5,3,22,1,13,14,15,11,21,12,19,2,4,8,10,7,18,17,6	0.039	0.0520	0.024
9,20,5,3,22,1,13,14,15,11,21,12,19,2,4,8,10,7,18,17,6,16	0.039	0.0520	0.024

In Figs. 5.3 (c) and (d), we show that attributes 3 and 4 lead to the lowest error rates (Table 5.27) in both SUD and Relief-F methods. Hence, the selected attribute subset for the Iris data set is $\{3, 4\}$ according to SUD and Relief-F, which is the same as the result based on our SCM method.

We obtained the same attribute ranking results and hence the same attribute subsets using our SCM with both bottom-up and top-down searches for the Monk3 data set (Table 5.28). Figs. 5.4(a) and (b) show that attributes 2, 4, and 5 should be selected for the Monk3 data set, which decreases the classification error rate from 0.0688 to 0.067, the number of inputs from 6 to 3, and the number of Gaussian hidden units from 19 to 11 (Table 5.16). In Figs. 5.4 (c) and (d), attributes 2, 4, and 5 should be selected according to both SUD and Relief-F methods, which is the same as the attribute subset obtained based on our SCM method.

Table 5.21. Attribute importance ranking for the Ionosphere data set using the SCM with different χ's obtained by bottom-up search.

χ	Ionosphere
$\chi = 0.0$	15 21 23 13 31 19 29 28 5 17 25 7 3 11 24 33 8 20 32 22 26 9 14 18 30 12 27 10 4 6 16 34 1 2
$\chi = 0.4$	5 3 7 15 31 21 29 23 13 25 19 33 17 9 28 11 8 14 24 20 22 12 32 18 26 30 27 10 6 4 16 34 2 1
$\chi = 0.5$	5 3 7 31 15 21 29 23 13 25 33 9 19 17 11 8 28 14 12 24 22 20 18 32 27 26 10 6 30 4 16 2 34 1
$\chi = 0.7$	5 3 2 7 31 15 29 21 23 13 33 25 19 11 17 8 14 12 28 27 18 6 22 10 16 24 20 32 4 26 30 34 1
$\chi = 1.0$	2 5 3 7 31 9 29 33 21 15 23 13 25 11 8 14 19 12 17 27 6 16 10 4 18 22 34 32 20 30 26 24 28 1

Table 5.22. Comparison for the Ionosphere data set between importance ranking results obtained by our SCM using bottom-up search when $\chi = 0.5$, the SUD and Relief-F methods.

Data set	Decreasing order of importance		
	SCM (bottom-up)	SUD	Relief-F
Ionosphere	5,3,7,31,15,21,29, 23,13,25,33,9,19, 17,11,8,28,14,12, 24,22,20,18,32,27, 26,10,6,30,4,16 ,2,34,1	13,15,11,9,7,17,19, 21,5,3,23,25,27, 29,31,33,10,4,6, 12,14,8,16,20,18, 22,28,26,24,30, 32,34,2,1	34,22,33,6,4,8, 16,14,21,9,27,15, 30,20,29,24,32,7, 12,18,10,11,3,5,28, 25,26,19,23,1,31, 13,17,2

Table 5.23. Comparison of the numbers of hidden units and classification errors before and after irrelevant attributes are removed according to the SCM ranking method for the Ionosphere data set. B: before removal, A: after removal.

Comparison	Data set	
	Ionosphere	
	B	A
Input attributes	34 attributes	first 21 attributes in the attribute queue
Number of hidden units	24	23
Classification error rate	7.13%	5.57%

For the Thyroid data set, the attribute ranking queue corresponding to our SCM with bottom-up search is $\{2, 3, 5, 4, 1\}$ and is $\{2, 5, 3, 4, 1\}$ with top-down search. Fig. 5.5 (a) and (b) show that, in both bottom-up and top-down searches, attributes 2, 3, and 5 are considered to be relevant for classification and are selected, which decreases the classification error rate from 0.0465 to

Table 5.24. Comparison of the numbers of hidden units and classification errors before and after irrelevant attributes are removed according to the SUD ranking method for the Ionosphere data set. B: before removal, A: after removal.

Comparison	Data set	
	Ionosphere	
	B	A
Input attributes	34 attributes	first 22 attributes in the attribute queue
Number of hidden units	24	22
Classification error rate	7.13%	6.86%

Table 5.25. Comparison of the numbers of hidden units and classification errors before and after irrelevant attributes are removed according to the Relief-F ranking method for the Ionosphere data set. B: before removal, A: after removal.

Comparison	Data set	
	Ionosphere	
	B	A
Input attributes	34 attributes	first 21 attributes in the attribute queue
Number of hidden units	24	20
Classification error rate	7.13%	5.43%

Table 5.26. Classification error rates for the Ionosphere data set with different subsets of attributes according to the importance ranking shown in Table 5.21 when $\chi = 0.5$. The attribute subset with the lowest validation error is highlighted in bold.

Number of attributes used	Error rate		
	Training set	Validation set	Test set
1	0.1706	0.1571	0.1571
5	0.1280	0.1286	0.10
10	0.1090	0.0715	0.10
15	0.0853	0.0715	0.1143
20	0.0853	0.0470	0.0857
21	**0.0711**	**0.0429**	**0.0571**
25	0.0711	0.0715	0.0571
30	0.0834	0.0571	0.0857
34	0.6350	0.857	0.0713

0.0233, the number of inputs from 5 to 3, and the number of Gaussian hidden units from 8 to 5 (Table 5.16). It is shown in Fig. 5.5(c) that attributes 4, 5, 3, and 2 are selected based on the ranking result of the Relief-F method.

Fig. 5.5(d) shows that attributes 4, 3, 1, and 2 should be selected according to SUD. The classification error rates of the test data set when the respective selected attribute subsets are used as inputs for RBF classifiers are 0.0233, 0.093, and 0.1163 for our SCM method, SUD, and Relief-F, respectively (Table 5.27). Hence, the attribute subset based on our SCM method is smaller with higher accuracy compared to the SUD and Relief-F methods.

For the Breast cancer data set, the attribute ranking queue corresponding to our SCM with bottom-up search is $\{2, 7, 3, 4, 9, 5, 1, 8, 6\}$ and is $\{7, 2, 3, 4, 9, 5, 8, 6, 1\}$ with top-down search. It is shown in Figs. 5.6(a)-(b) that, in both bottom-up and top-down searches, attributes 2, 3, and 7 are considered to be important for classification and are then selected, which decreases the classification error rate from 0.0146 to 0.0073, the number of inputs from 9 to 3, and the number of hidden units of the RBF neural network from 11 to 5 (Table 5.16). Fig. 5.6(c) shows that the attribute subset including the first five attributes (attributes 6, 2, 3, 7, and 5) in the Relief-F attribute ranking queue leads to the lowest classification error rates. According to the classification results shown in Fig. 5.6(d) based on the ranking result of the SUD method, attributes 1, 7, 3, 2, and 5 should be selected because the subset leads to the lowest error rates. The classification error rates of the test data set when the selected attribute subsets are used as inputs for RBF classifiers are 0.0073, 0.0146, and 0.0073 for our SCM, SUD, and Relief-F, respectively (Table 5.27). Hence, the attribute subset based on our SCM method is the smallest with the highest classification accuracy.

For the Mushroom data set, the attribute ranking queues corresponding to our SCM with bottom-up search and top-down searches are the same: $\{9, 20, 5, 3, 22, 1, 13, 14, 15, 11, 21, 12, 19, 2, 4, 8, 10, 7, 18, 17, 6, 16\}$. In the Mushroom data set, the 16th attribute has the same value for all samples. It is clear that it is an irrelevant attribute and it will not play any role in classification. It is shown in Table 5.18 that our method and the SUD method rank this attribute as the least important attribute. It is shown in Figs. 5.7(a)-(b) that, in both bottom-up and top-down searches, attributes 9, 20, and 5 are considered to be important for classification and are then selected, the number of inputs reduces from 22 to 3, and the number of hidden units of the RBF neural network reduces from 29 to 18 (Table 5.19). Fig. 5.7(c) shows that the attribute subset including the first two attributes (attributes 5 and 20) in the Relief-F attribute ranking queue leads to the lowest classification error rates. According to the classification results shown in Fig. 5.7(d) based on the ranking result of the SUD method, attributes 9, 3, 20, 22, and 5 should be selected because the subset leads to the lowest error rates.

5.8 Summary

In this chapter, a novel separability-correlation measure (SCM) is described to rank the importance of attributes. SCM is composed of two measures:

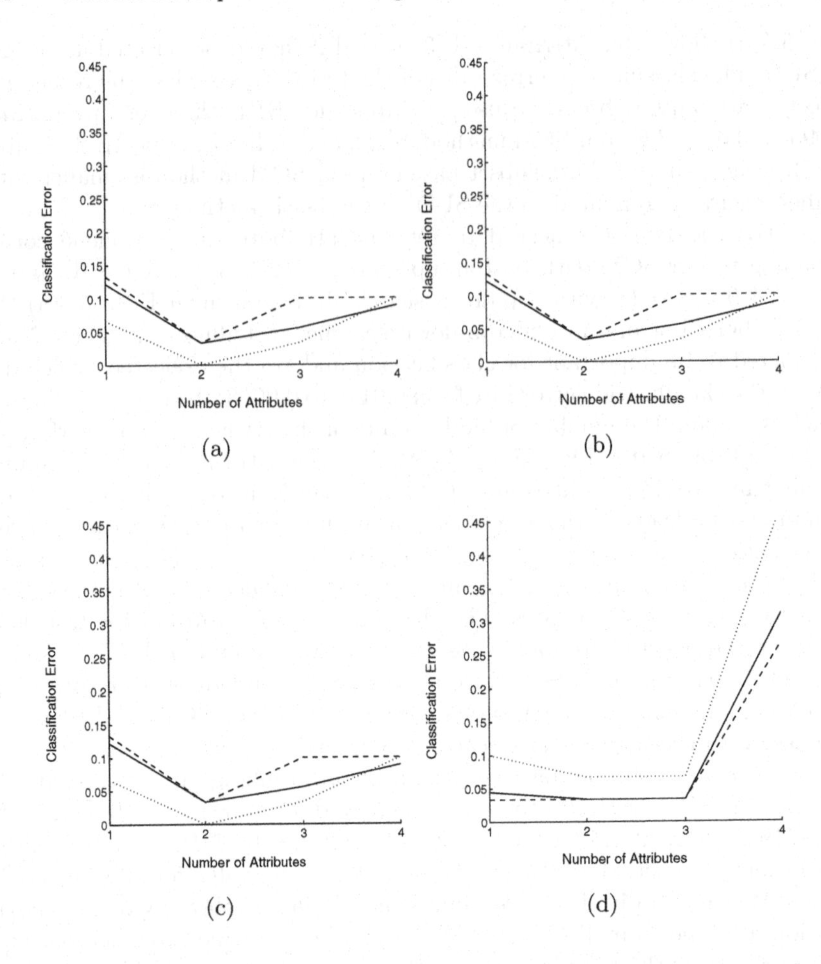

Fig. 5.3. Classification error rates of the Iris data set for different numbers of attributes used according to attribute ranking results obtained from (a) SCM with bottom-up search, (b) SCM with top-down search, (c) Relief-F, and (d) SUD. Solid line: the training data set; dotted line: the validation data set; dashed line: the test data set. (© 2005 IEEE) We thank the IEEE for allowing the reproduction of this figure, first appeared in [107].

the class-separability measure and the attribute-class correlation measure. Though the class-separability measure, i.e., the ratio of the interclass distance and the intraclass distance, delivers information for the discriminatory capability of attributes, this measure does not always work well alone. The correlation between attributes and class labels is calculated as another measure for the discriminatory ability of attributes, which measures the influence of the change of attributes on the change of class labels for patterns. We use

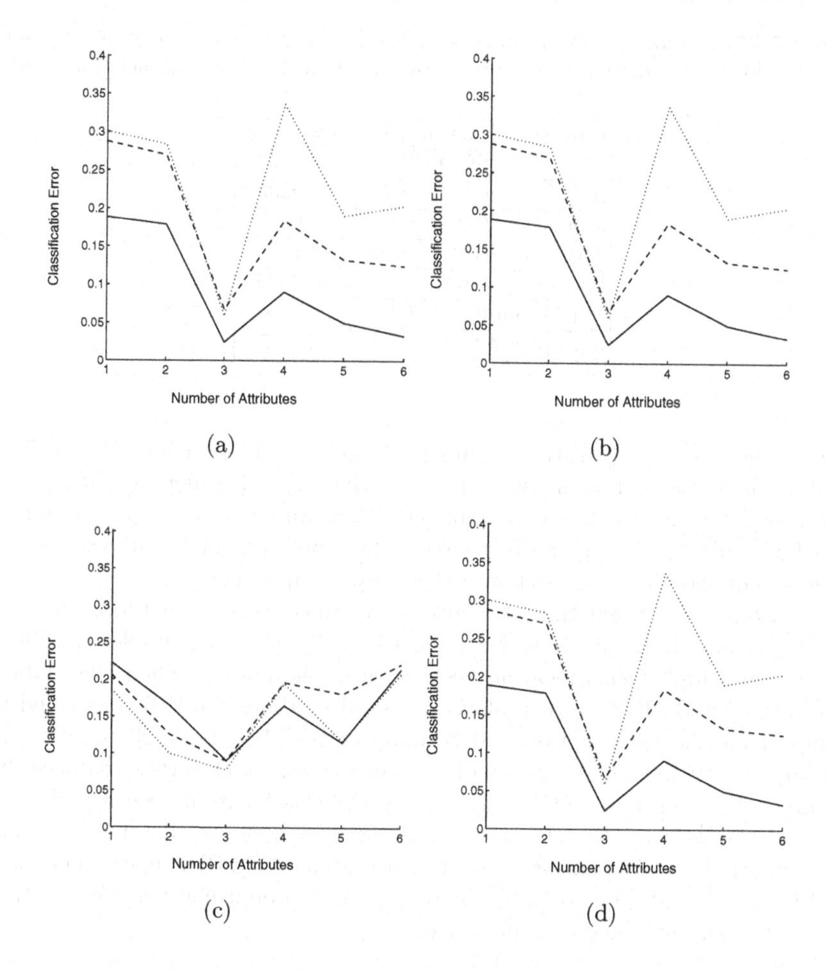

Fig. 5.4. Same as Fig. 5.3, Monk3 data set. (© 2005 IEEE) We thank the IEEE for allowing the reproduction of this figure, first appeared in [107].

a weighting parameter that leads to the highest classification accuracy. Attribute importance ranking results from SUD [71] and Relief-F [180] methods are shown.

In this chapter, data dimensionality reduction is also carried out in order to improve classification performance and to reduce the number of attributes as well as the complexity of the RBF neural network.

According to the ranking results obtained by the SCM method, different attribute subsets are used as inputs to RBF classifiers. The attribute subsets with the lowest classification error rates and the least numbers of attributes are selected. With bottom-up and top-down searches, the same selected attribute subsets are eventually obtained in the five benchmark data sets tested.

Table 5.27. Comparison between classification error rates of the testing data sets with the best attribute subsets obtained by our SCM, SUD, and Relief-F methods.

Data set	classification error rates		
	SCM	SUD	Relief-F
Iris	0.0333	0.0333	0.0333
Monk3	0.067	0.067	0.09
Thyroid	0.0233	0.093	0.1163
Breast	0.0073	0.0146	0.0073
Mushroom	0.012	0.02	0.008
Ionosphere	0.0279	0.031	0.047

The ranking operation is independent of induction algorithms. The ranking is not linked with the training of the RBF classifier, which reduces the computational cost. By ranking attribute importance, fewer candidate attribute subsets need to be inspected. Thus, our method combines the advantages of the filter approach and the wrapper approach.

Compared to existing attribute importance ranking methods, such as SUD [71] and Relief-F [180] methods, the SCM leads to smaller attribute subsets and higher classification accuracies in simulations when class labels are available. We have also employed a useful modification [107] described in Chap. 4 for the construction and training of the RBF network by allowing for large overlaps among clusters of the same class, which further reduces the number of hidden units while maintaining the classification accuracy. Experimental results show that the methods described here are effective in reducing the attribute size, the structural complexity of the RBF neural network, and the classification error rates. Though it is a suboptimal feature selection method, high performance is obtained.

Table 5.28. Comparison between importance ranking results obtained by our SCM using bottom-up and top-down searches when $\chi = 0.4$, the SUD and Relief-F methods. (© 2005 IEEE) We thank the IEEE for allowing the reproduction of this table, first appeared in [107].

Data set	Decreasing order of importance			
	SCM (bottom-up)	SCM (top-down)	SUD	Relief-F
Iris	4,3,1,2	4,3,1,2	3,4,1,2	4,3,1,2
Monk3	5,2,4,1,6,3	5,2,4,1,6,3	5,2,4,1,6,3	2,5,4,3,6,1
Thyroid	2,3,5,4,1	2,5,3,4,1	4,5,3,2,1	4,3,1,2,5
Breast	2,7,3,4,9,5,1,8,6	7,2,3,4,9,5,8,6,1	1,7,3,2,5,6,4,8,9	6,2,3,7,5,1,4,8,9

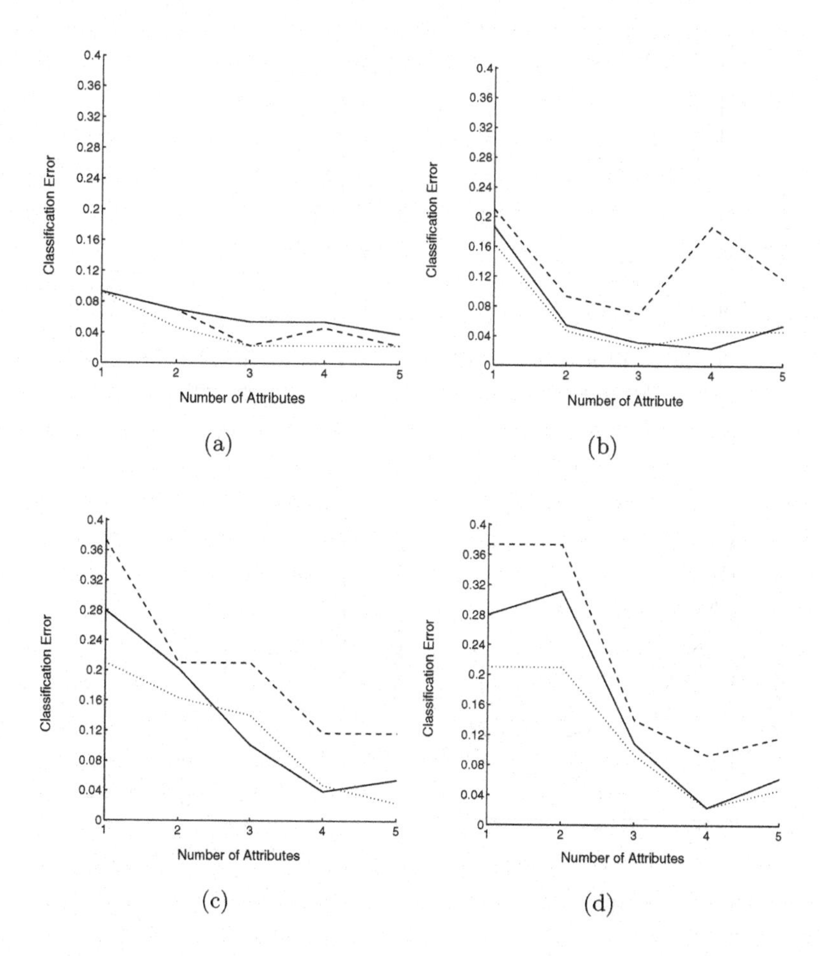

Fig. 5.5. Same as Fig. 5.3, the Thyroid data set. (© 2005 IEEE) We thank the IEEE for allowing the reproduction of this figure, first appeared in [107].

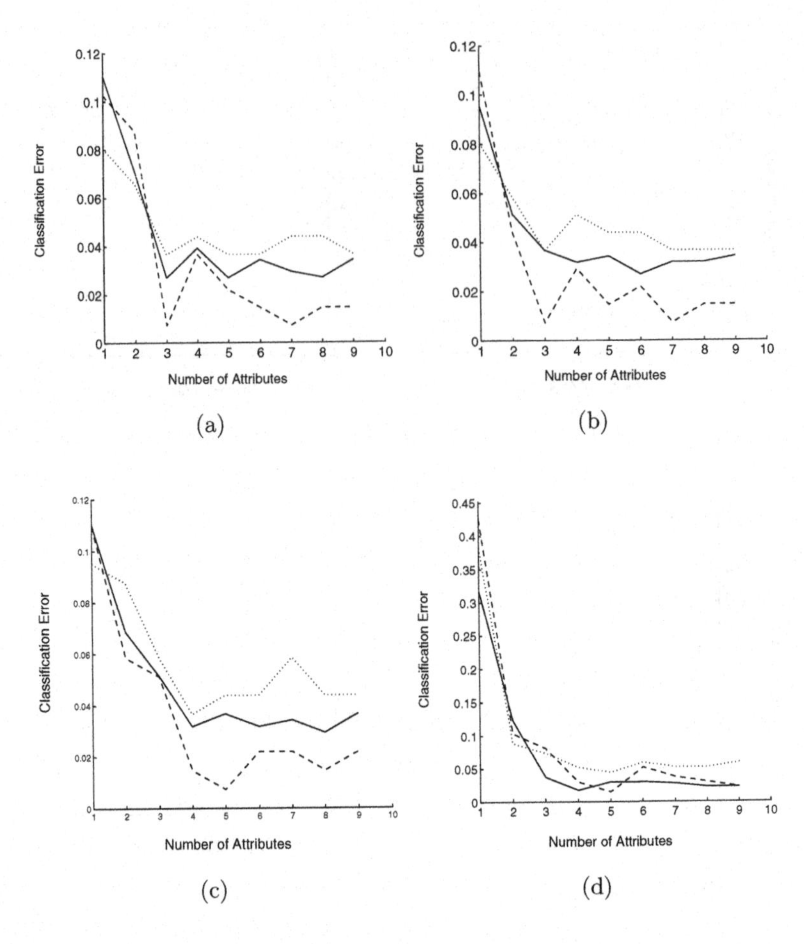

Fig. 5.6. Same as Fig. 5.3, the Breast cancer data set (note that the scale for classification error rates in (d) is different from those in (a)-(c)). (© 2005 IEEE) We thank the IEEE for allowing the reproduction of this figure, first appeared in [107].

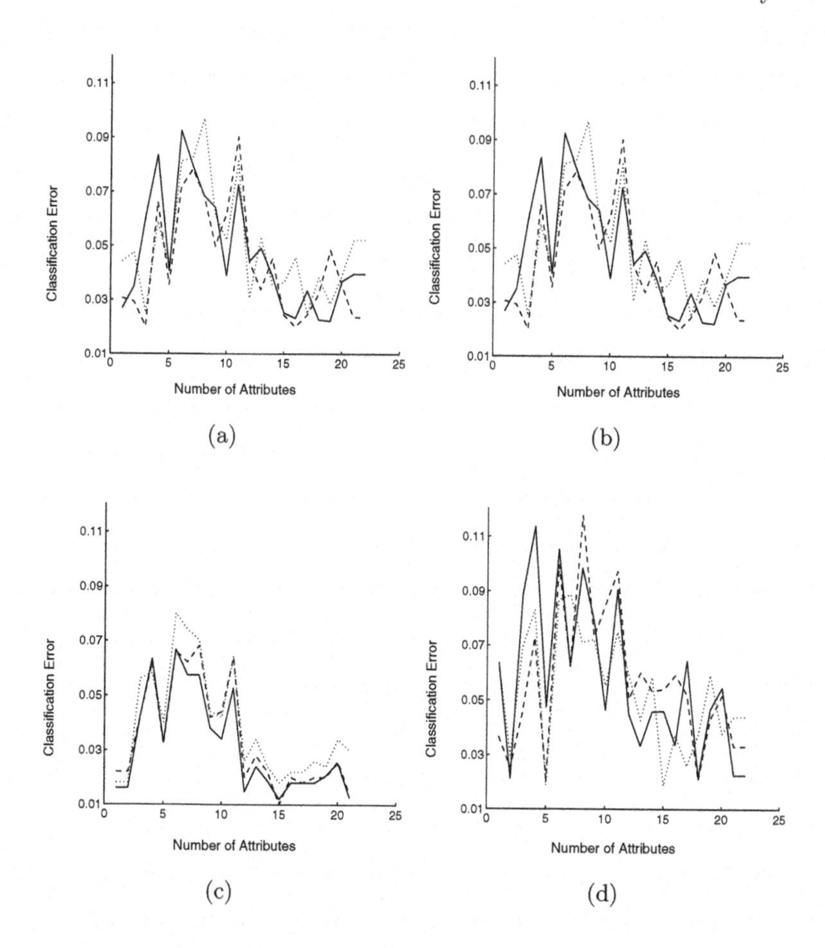

Fig. 5.7. Same as Fig. 5.3, the Mushroom data set. (© 2005 IEEE) We thank the IEEE for allowing the reproduction of this figure, first appeared in [107].

6

Genetic Algorithms for Class-Dependent Feature Selection

6.1 Introduction

Removal of redundant or irrelevant attributes from data sets can facilitate practical applications in improving speed and relieving memory constraints. Data dimensionality reduction (DDR) can reduce the computation burden in semi-automated or automated processes, for example when constructing a radial basis function (RBF) neural network to classify data.

Many algorithms have been developed for DDR. DDR can be categorized from different aspects. As stated in previous chapters, DDR can be classified into feature selection and feature extraction based on the origins of resultant features. According to the relationship of DDR methods with induction systems, DDR techniques can be categorized as the filter approach and the wrapper approach. There is another way to categorize DDR techniques, i.e., class-*independent* feature selection, in which features selected are common to all classes, and class-*dependent* feature selection, in which different feature sets are selected for different classes.

In class-independent feature selection [18][27][166][199][207][209], all features selected are assumed to play equal roles in discriminating each class from the others, which hides the possibility that different groups of features may have different capabilities in discriminating classes.

Some class-independent feature selection techniques were proposed based on genetic algorithms (GAs) [32][44][112][185][254], which are popular searching algorithms. In GA feature selection techniques, each chromosome in the population pool represents a feature mask [32][44][185][254]. Assume that there are n original features in a data set. n bits are needed in a chromosome to represent the n features. The kth bit of a chromosome indicates the presence or absence of the kth feature. Usually, the feature is present if the bit is 1, and is absent if the bit is 0. The classification accuracy of a classifier is used as the fitness function in GAs. Fung *et al.* [112] proposed to use fuzzy GAs (FGAs) to select class-independent features. Instead of selecting or rejecting a feature completely, Fung *et al.* introduced in-between cases for determining the importance of a feature, i.e., the fitness of a feature is between 0 and 1 according to its importance in FGAs. In the feature selection based on FGAs, features are considered generally to all classes. In [32][112], the fitness evaluator was a nearest-neighbor classifier. Chaikla and Qi [44] also chose the nearest-neighbor classifier together with multiple correlation as the fitness function of GAs to select class-independent features. Raymer *et al.* [254] encoded the number of nearest-neighbor classifiers into a chromosome together with the features. The classification accuracy was also used as the evaluation function. In [185], the classification result was determined by the vote of several different classifiers, i.e., the logistic classifier (LOG), the linear discriminant classifier (LDC), and the quadratic discriminant classifier (QDC), etc.

Consider a feature vector X, $X = \{x_j \mid j = 1, 2, ..., L\}$, where x_j represents the jth feature. In a classification task, a training sample $X_i = \{x_{ij} \mid j = 1, 2, ..., L\}$ is given with its class label y_i. Class-dependent feature selection is to select the best subset $Z_k = \{Z_{kj} \mid k = 1, 2, ..., M, j = 1, 2, ..., l_k, Z_{kj} \in X\}$ with $l_k \leq L$ for each class k. M is the number of classes in a data set. Class-dependent feature selection techniques are developed based on the fact that a feature may have a different power in discriminating classes, and thus a subset of features may be selected to discriminate one class from other classes.

In [232][233], Oh *et al.* proposed a class-dependent feature selection method to improve pattern recognition performance. Based on their observation, class-dependent features played important roles in class separation. Multiple MLP classifiers [232][233] were constructed with inputs of the class-dependent features. M MLP classifiers were trained corresponding to a data set with M classes, which led to high cost in computation. A feature subset for the ith class is evaluated by class separation S^{cc}. The separation between class w_i and class w_j was defined by Oh *et al.* as:

$$S^{cc}(w_i, w_j, \mathbf{X}) = \int_{RL} |P_n^{w_i}(X) - P_n^{w_j}(X)| dx, \tag{6.1}$$

where RL is a L-dimensional real space. $P_n^{w_i}(X)$ and $P_n^{w_j}(X)$ are estimated distributions for class w_i and class w_j, respectively. To select the feature subset for class p, all the other classes are combined and treated as the other class q. Each attribute x_k $(0 < k \leq L)$ is evaluated by class separation $S^{cc}(w_p, w_q, x_k)$ and put in a ranking queue according to the magnitude of its corresponding S^{cc}. The larger $S^{cc}(w_p, w_q, x_k)$ is, the more important the attribute x_k is considered to be. Then the L attribute subsets, where the jth attribute subset g_j comprises the first j most important attributes in the ranking queue, are evaluated with $S^{cc}(w_p, w_q, g_k)$ to obtain the best class-dependent feature subset with the largest S^{cc} for each class.

Bailey [14] defined class-dependent features as *features whose discrimination ability varies significantly depending on the classes which are to be discriminated.* Strong and weak class-dependence are defined when discussing class-dependent features [14]:

- Weak class-dependence: the feature has an observable value for each class. The feature and the class variable are conditionally independent given a set of classes.
- Strong class-dependence: when conditioned on the class variable, the feature may not have an observable value.

Strong or weak class-dependence could be considered in some particular domain, such as speech signals. Some measurements for speaking or writing manners are usually conditioned on the individual person (a class label in the data), which may not be observable on another person.

Besides class-dependent feature selection, class-specific feature extraction is also used in the literature to generate features. Baggenstoss [11][12] presented a new approach to deliver feature subsets separately for each class of data. Class-specific features are transformed from the original data space in order to transform original high dimensionality into a much lower dimensionality. This approach facilitated the construction of classifiers based on the joint probability density function (PDF) of class-specific features which are transformed from original features. In a probabilistic classifier, it is very complex to calculate the PDF when the feature dimensionality exceeds five. The class-specific method [13] is composed of two steps. First, original features are transformed into a low-dimensional feature space, and then the PDF is estimated in the new feature space. Second, the feature PDFs are projected back to the PDFs in the original space.

In Kay's work [169], a sufficient statistic for each class is considered separately. Class-specific features are generated for facilitating the estimation of the probability density function (PDF) when constructing optimal classifiers.

In [234], a non-linear class-dependent feature transformation is described. The transformation of features is done by minimizing an estimate of the relative entropy between actual conditional likelihood and its approximation. The new feature vector is calculated using the current symplectic transformation parameters in each iteration, and the maximum likelihood estimates of the HMM (Hidden Markov Model) parameters are calculated subsequently. After the HMM and symplectic parameters have converged locally, the class-dependent feature subsets are transformed by the symplectic map. The maximum likelihood estimates of symplectic map parameters are obtained by the conjugate gradient algorithm. It is noted that the acoustic features generated for a certain class by the above class-dependent feature method may not be observable for other classes.

In this chapter, we describe a method that we proposed earlier [104] for selecting class-dependent features and constructing a novel RBF classifier based on class-dependent features. For different groups of hidden units corresponding to different classes in RBF neural networks, different feature subsets are selected as inputs. GAs are used to search for the optimal feature masks for all classes. In contrast to Oh *et al.* [232][233], only a single such RBF network, rather than multiple MLPs, is required for a multi-class problem when selecting class-dependent features.

6.2 The Conventional RBF Classifier

In conventional RBF neural networks, all attributes are used as inputs. And no weights need to be tuned between the input layer and the hidden layer, i.e., default weights are 1's. The activation of a hidden unit is determined by the distance between the input vector and the center vector of the hidden unit. A hidden unit is a cluster, which is active only for a subset of patterns. As stated in Chap. 4, most patterns in the subset have the same class label as the initial center of the hidden unit. In our RBF neural network training algorithm, the initial centers are randomly selected from the training set. The weights connecting hidden units and output units will show the relationship between hidden units and classes. Let us consider a weight matrix W:

$$W = \begin{pmatrix} w_{11} & w_{12} & \cdots & w_{1M} \\ w_{21} & w_{22} & \cdots & w_{2M} \\ \cdots & \cdots & \cdots & \cdots \\ w_{m1} & w_{m2} & \cdots & w_{mM} \end{pmatrix} , \quad (6.2)$$

where w_{ik} $(i = 1, ..., m,\ k = 1, ..., M)$ is the weight which connects hidden neuron i with output neuron k. m is the number of hidden neurons and M is the number of class labels, which is the same as the number of output neurons in the RBF neural network.

For example, the corresponding matrix of the Thyroid data set is:

$$W = \begin{pmatrix} 0.91 & -0.08 & -0.15 \\ 0 & 1.74 & -0.380 \\ 0.97 & -0.18 & -0.2 \\ -0.10 & -0.04 & 1.61 \\ -0.14 & 0.87 & -0.05 \\ -0.09 & 0.93 & -0.10 \end{pmatrix} .$$

Here six hidden units are constructed for three classes. As shown in W, the first and the third hidden units mainly serve the first output neuron (class 1), since only w_{11} and w_{31} are significant in row 1 and row 3, respectively. Similarly, the second, the fifth, and the sixth hidden units mainly serve class 2, and the fourth hidden unit mainly serves class 3.

The hidden units serving a class are not generated sequentially during training. But, we can group them to show that every class corresponds to a subset of hidden units as in Fig. 6.1.

Thus, class-dependent features can be selected through RBF neural networks, i.e., a subset of features for a certain class are the inputs to only

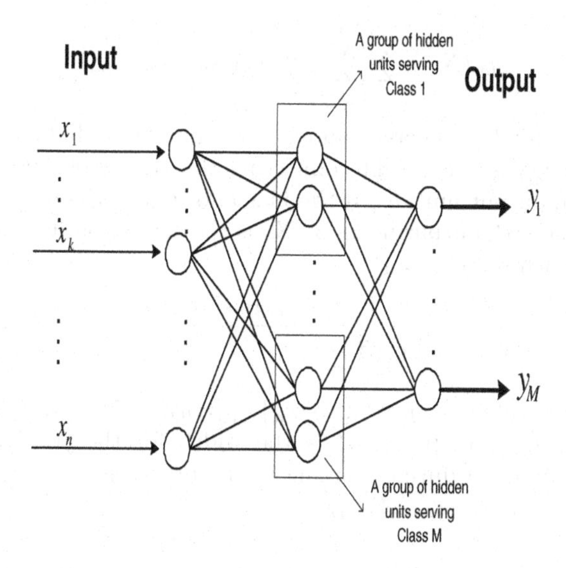

Fig. 6.1. Architecture of a conventional RBF neural network.

the hidden units which mainly serve the class. A novel RBF classifier is constructed based on this idea. We will show it in the next section.

6.3 Constructing an RBF with Class-Dependent Features

For a classifier, classification accuracy or the capability for separating classes is a main evaluation criterion of its performance. The best feature subsets for each class can be obtained by evaluating every combination of feature subsets together with the resultant classification accuracy. However, in practice, it is too computationally expensive to implement. We employ GAs for searching for optimal class-dependent feature subsets.

6.3.1 Architecture of a Novel RBF Classifier

It is observed that the hidden neurons in an RBF network may be grouped according to classes. That is, if most of the patterns in the cluster represented by a hidden neuron belong to class i, we say that this hidden neuron belongs to the group for class i (Fig. 6.2). We add a class-dependent feature mask for each group of hidden neurons. $\{x_1^i, x_2^i, ..., x_{k_i}^i\}$ (k_i is the number of class-dependent features for class i) are the features selected for discriminating class i from other classes.

The mth output of the network is as follows:

$$y_m(\mathbf{X}) = \sum_{i=1}^{M} \sum_{j=1}^{k_i} w_{mj}^i \o_j^i(\mathbf{X}) + w_{m0} b_m, \tag{6.3}$$

where M is the number of classes and k_i is the number of hidden units serving the ith class. w_{mj}^i is the weight connecting the jth hidden unit of the mth class to the ith output unit. b_m is the bias and w_{m0} is the weight connecting the bias and the mth output node. $\o_j^i(\mathbf{X})$ is the activation function of the jth hidden unit which serves class i:

$$\o_j^i(\mathbf{X}) = \mathrm{e}^{\frac{-||\mathbf{X}^i - \mathbf{C}_j^i||^2}{2\sigma_j^{i\,2}}}. \tag{6.4}$$

Here $\mathbf{X}^i = \{g_1^i x_1, g_2^i x_2, ..., g_k^i x_k, ..., g_n^i x_n\}$. $\{g_1^i, g_2^i, ..., g_k^i, ..., g_n^i\}$ is the feature mask for class i. $g_k^i \in \{0, 1\}$. σ_j^i is the width for the jth hidden unit of class i and is obtained during training in the presence of the feature masks. $\mathbf{C}_j^i = \{g_1^i c_1, g_2^i c_2, ..., g_k^i c_k, ..., g_n^i c_n\}$.

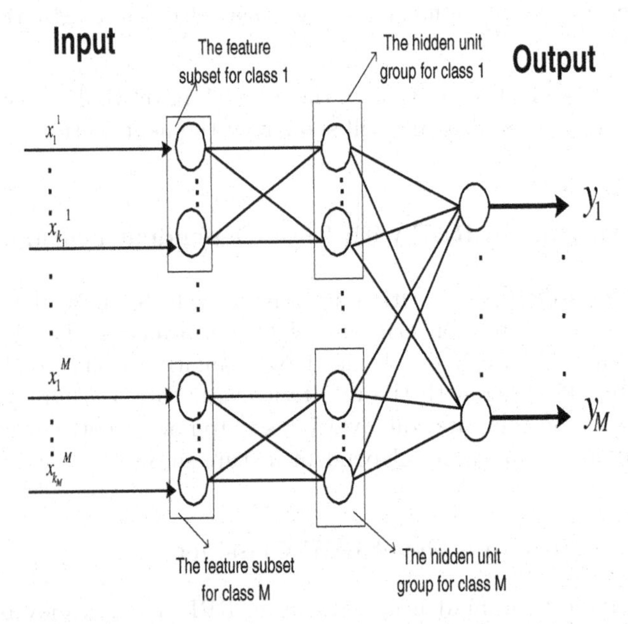

Fig. 6.2. Architecture of a new RBF neural network with class-dependent feature masks. (© 2005 IEEE) We thank the IEEE for allowing the reproduction of this figure, first appeared in [104].

6.4 Encoding Feature Masks Using GAs

A GA exhibits powerful efficiency in solving many complex problems. It can search for optimal solutions based on its fitness evaluation and offspring generation strategies. Usually, a binary string (an individual in the population pool) is used to represent a solution for a problem. Each individual is evaluated by the defined fitness function. The operators of a GA, such as selection, crossover, and mutation, are used for producing offspring. Parents with higher fitness score are given high probabilities to generate offspring.

GAs have been used widely in the literature [32][44][112][185][254]. Here, we use GAs for determining class-dependent feature masks. Suppose that n is the total number of the original features and M is the number of classes. A binary string representing a possible solution in a GA is shown in Fig. 6.3. The length of each individual is nM bits. A chromosome G is represented as follows:

$$G = \{(g_1^1, ..., g_i^1, ..., g_n^1), ..., (g_1^k, ..., g_i^k, ..., g_n^k), ..., (g_1^M, ..., g_i^M, ..., g_n^M)\}. \quad (6.5)$$

Here $g_i^k \in \{0, 1\}$, $k = 1, 2, ..., M$, and $i = 1, 2, ..., n$.

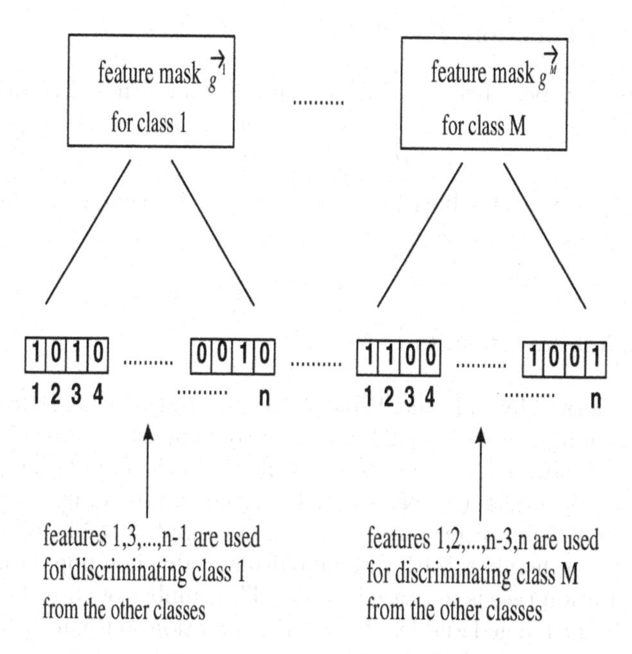

Fig. 6.3. An encoding string presenting an exemplary solution. (© 2005 IEEE) We thank the IEEE for allowing the reproduction of this figure, first appeared in [106].

6.4.1 Crossover and Mutation

We use the roulette wheel selection to select chromosomes in each generation. In the roulette wheel selection, the selection probability is proportional to each chromosome's fitness. Two-point crossover is used. Two points are randomly located in each of the two parents. The two parts of the parent chromosomes between the two pairs of points are then exchanged to generate new offspring. The probability of crossover is chosen to be around 80% in this book.

Mutation can prevent fixation at some particular loci. A locus in the parent chromosome is selected randomly and the bit at the position is replaced, i.e., if the original bit is 0, it is replaced by 1, and vice versa. Usually, the mutation rate is relatively small to avoid too much variation. However, at later generations, the number of identical members increases, which leads to a stagnant state. In order to break stagnant states to search for optimal results, we use a dynamic mutation rate, i.e., if the number of identical members in a population exceeds a certain percentage, the mutation rate is increased by a certain amount.

6.4.2 Fitness Function

For our purpose here, we choose the following fitness function:

$$F(G) = 1 - E_v(G), \tag{6.6}$$

where $E_v(G)$ is the classification error rate of the validation data set for chromosome G.

6.5 Experimental Results

The Glass, Thyroid, and Wine data sets from the UCI Repository of Machine Learning Databases [223] are used to demonstrate our algorithm.

The GA parameters are chosen as follows. In the population pool, there are nM (M is the number of classes, and n is the number of features) chromosomes. The initial mutation rate is 40%. If the number of identical members or the number of members with the same fitness value in a population exceeds 25%, the mutation rate is increased by 1%. The number of elite chromosomes, which remain unchanged and live from one generation to the next, is two. The generation number is 50. In GAs, chromosomes in an initial pool are generated randomly. The parameters are not sensitive to the data sets. We determine the parameters empirically according to [8][44][351] etc.

6.5.1 Glass Data Set

The collection of the Glass data set was for the study of different types of glass, which was motivated by criminological investigations. At the scene of a crime, the glass left can be used as evidence if it is correctly identified The Glass data set contains 214 cases. There are nine attributes and six classes in the Glass data set. 114 patterns are for training, 50 for validation, and 50 for testing.

The attributes are RI (refractive index), Na (sodium), Mg (magnesium), Al (aluminum), Si (silicon), K (potassium), Ca (calcium), Ba (barium), and Fe (iron). The six classes of the Glass data set are building-windows-float-processed, building-windows-non-float-processed, vehicle-windows-float-processed, containers, tableware, and headlamps.

The error rates in our experiments are 16.41% for the training set, 20.93% for the validation set, and 23.26% for the test set. For class 1, four features are involved for discriminating it from other classes, i.e., features $\{3, 4, 6, 8\}$ obtained by the GA are used as the inputs to the hidden units of class 1 (Table 6.1). The feature subset $\{1, 2, 3, 4, 5, 7, 9\}$ is used as the input to the hidden units of class 2. The feature subset $\{1, 2, 5, 6, 8\}$ is used to discriminate class 3 from other classes. For class 4, the corresponding feature subset is $\{2, 3, , 5, 7\}$. The feature subset $\{4, 5, 6, 8, 9\}$ is the input for class 5, and the feature subset $\{1, 2, 4\}$ is the input for class 6. Thus, the average number of features used for each class is 4.7 compared to the original nine features.

Table 6.1. The feature mask found by GAs for the Glass data set. (© 2005 IEEE) We thank the IEEE for allowing the reproduction of this table, first appeared in [104].

Classes	Feature masks
Class 1	0 0 1 1 0 1 0 1 0
Class 2	1 1 1 1 1 0 1 0 1
Class 3	1 1 0 0 1 1 0 1 0
Class 4	0 1 1 0 1 0 1 0 0
Class 5	0 0 0 1 1 1 0 1 1
Class 6	1 1 0 1 0 0 0 0 0

Our experimental result is compared with other methods' results [185] in Table 6.2. In the first line of Table 6.2, the classification result of the LOG classifier [82][185] using all the features (no feature selection) is shown. By sequential backward selection (SBS) [76][185], 6.9 class-independent features on average are obtained for all classes. The LOG classification result [185] based on selecting class-independent features using GAs is shown in the third line of

Table 6.2. Accuracy comparison with other methods for the Glass data set. () We thank the IEEE for allowing the reproduction of this table, first appeared in [104].

Method	Training error	Testing error	Number of features
LOG without feature selection	28.14%	39.54%	9
SBS	27.62%	37.39%	6.9
LOG with class-independent feature selection	27.52%	40.39%	7.7
Multiple classifiers Version 1	27.62%	39.26%	Unknown
Multiple classifiers Version 2	25.35%	34.55%	Unknown
Our method	16.41%	23.26%	4.7

Table 6.2. There are two versions of another algorithm involving multiple classifiers [185], i.e., the logistic classifier (LOG), the linear discriminant classifier (LDC), and the quadratic discriminant classifier (QDC), etc. In multiple classifiers (version 1) [185], a GA was used for class-independent feature selection and only features were encoded in the GA. The fitness of each chromosome was determined by a vote of the classifiers. In multiple classifiers (version 2), the classifier types were also encoded in the GA, i.e., each individual classifier can be chosen from the three classifiers mentioned above. The comparison shows that a better classification accuracy is obtained with a fewer number of features by our method.

6.5.2 Thyroid Data Set

The detailed information about the Thyroid data set is shown in Chap. 4. 115 patterns of the Thyroid data set are used for training, 50 are for validation, and 50 are for testing.

With class-dependent features, the classification error rates are 2.84% for the training set, 2.33% for the validation set, and 4.65% for the test set. Without feature masks, the classification error rates are 3.88% for the training set, 3.88% for the validation set, and 4.65% for the test set.

It is shown in the feature masks (Table 6.3) that feature 1 does not play any important role in discriminating the classes. For class 3, feature 2 is used to discriminate it from other classes. Features 2 and 3 are used to discriminate class 2 from other classes. Features $\{2, 3, 4, 5\}$ are used for discriminate class 1 from other classes. The average number of features used for each class is 2.33, compared to the original five features.

Table 6.3. The feature mask found by GAs for the Thyroid data set. (© 2005 IEEE) We thank the IEEE for allowing the reproduction of this table, first appeared in [104].

Classes	Feature masks
Class 1	0 1 1 1 1
Class 2	0 1 1 0 0
Class 3	0 1 0 0 0

6.5.3 Wine Data Set

The Wine data set was obtained from chemical analysis of wines produced in the same region of Italy but derived from three different cultivars. There are 13 attributes and 178 patterns in the Wine data set. 106 patterns are for training, 36 for validation, and 36 for testing. There are three classes corresponding to the three different cultivars.

With class-dependent features, the classification error rates are 2.83% for the training set, 0% for the validation set, and 2.78% for the test set. Without feature masks, the classification error rates are 3.77% for the training set, 2.78% for the validation set, and 2.78% for the test set.

It is shown in the feature masks (Table 6.4) that features 1 and 8 do not play any important role in discriminating the classes. For class 1, the feature subset $\{2, 4, 5, 6, 7, 9, 11, 12\}$ can discriminate it from other classes. Features 3, 4, 5, 6, 7, 10, 11, 12, and 13 are used to discriminate class 2 from other classes. The feature subset $\{3, 4, 11, 12, 13\}$ is used to discriminate class 3 from other classes. The average number of features used for each class is seven, compared to the original 13 features.

Table 6.4. The feature mask found by GAs for the Wine data set.

Classes	Feature masks
Class 1	0 1 0 1 1 1 1 0 1 0 1 1 0
Class 2	0 0 1 1 1 1 1 0 0 1 1 1 1
Class 3	0 0 1 1 0 0 0 0 0 0 1 1 1

6.6 Summary

In this chapter, we have selected class-dependent features for each class and described a novel RBF classifier with class-dependent features which are selected by GAs based on RBF classification performance. The feature subset is selected for each class individually based on its ability in discriminating the class from other classes, which brings out the relationship between the feature subset and the class concerned. The Glass, Thyroid, and Wine data sets are used to demonstrate the algorithm. Experimental results show that the algorithm is effective in reducing the number of feature inputs and improving classification accuracies simultaneously.

DDR is often the first step for data mining tasks. The class-dependent feature selection results obtained above provide new information for analyzing the relationship between features and classes. The reduction in dimensionality can lead to compact rules in the rule extraction task. Extracting rules based on the classification results obtained above will be shown in a latter chapter.

7

Rule Extraction from RBF Neural Networks

In this chapter, we first review rule extraction techniques. Next, a type of data mining systems, i.e., the rule extraction system, is discussed from the viewpoint of its components. Then, four decompositional rule extraction methods based on RBF neural networks are described. The first rule extraction method extracts rules from trained RBF neural networks through a GA: the GA is used to determine the rule premises [100]. The second extracts rules from trained RBF neural networks by a gradient descent method[101][333]. In the third rule extraction method, we extract rules based on the result of data dimensionality reduction using the gradient descent method [102]. The fourth rule extraction method utilizes the results of class-dependent feature selection to extract rules [106].

7.1 Introduction

A major problem in data mining using neural networks is that knowledge hidden in a trained neural network is not comprehensible to humans. Linguistic rule extraction [28][102] [269][287][317][324] aims at solving this problem.

Rule extraction can facilitate data mining in many aspects:

- Increase perceptibility and help human beings better understand decisions of learning models. This advantage of rule extraction is extremely helpful in medical diagnosis.
- Refine initial domain knowledge. Irrelevant or redundant attributes tend to be absent in extracted rules. In future data collections, labor cost can be reduced by skipping redundant or irrelevant attributes.
- Explain data concepts by linguistic rules to clients.
- Find active attributes in decision making. Many attributes may play roles in decision making. However, some attributes may be more active compared to others. Learning models usually are opaque in identifying active attributes. Rule extraction provides a solution in this problem.

In the literature, extracted rules are mainly evaluated based on two criteria [115][146]:

- rule accuracy,
- rule complexity.

An additional criterion is 'the fidelity' [286]. 'The fidelity' is defined as the ratio between correctly classified outputs from rule extraction and correctly classified outputs from the corresponding classifier. 'The fidelity' reflects how well a rule extraction method matches its corresponding classifier.

The objective of rule extraction is to obtain a comprehensible description of the data, rather than a description of the network structure in most cases. Based on this, we conclude that the fidelity measurement is not always necessary for evaluating rule extraction, especially when models based on which rules are extracted do not have high learning accuracy. High fidelity under this condition does not correspond to high performance in rules extracted, but to high mapping accuracy from rules extracted to learning models. In some cases, the accuracy of the extracted rules might be higher than the accuracy of the learning models, i.e., high consistency between a data concept and rules extracted is obtained rather than high consistency between rules extracted and learning models. Zhou [362] had questioned whether rule extraction is implemented using neural networks or for neural networks, i.e., to compensate for neural network 'black-box' behavior. And, Zhou [362] also concluded that rule extraction using neural networks and rule extraction for neural networks are two different tasks, and different criteria should be used for evaluating rule extraction techniques for the two tasks.

Given a data set, learning models from artificial intelligence and machine learning are employed to abstract essences of the data by training the models on a set of training patterns. Rule extraction is considered as a procedure to discover hidden information from data sets and represent it in explicit rules, which are relatively easy to understand, and potentially provoke new ideas in further data analysis.

For a data set with tens or hundreds of attributes and thousands of data patterns, it is hard to identify the roles of the attributes in classifying new patterns without any aid from learning models. Neural networks can be trained on these training samples to abstract essences and store the learned essential knowledge as parameters in the network. However, though essential knowledge has been captured and embedded in the trained neural network, humans cannot tell exactly why a new pattern is classified to a class, which is sometimes referred to as 'black-box' characteristics of neural networks. In the medical domain, a disjunctive explanation given as 'If medical measurement A is a_1, and medical measurement B is b_1,..., then conclusion.' is preferable to a complex mathematical decision function hidden in neural networks.

Rule extraction from neural networks has been an active research topic in recent years. In early rule extraction work, Gallant [116] used trained neural networks to develop an expert-system engine and interpret the knowledge embedded in neural network models by IF–THEN rules. More than a decade passed. The capability of rule extraction [1][139][288][301] had been shown for delivering comprehensible descriptions of data concepts from complex machine learning models.

Rule extraction techniques are usually based on machine learning methods such as neural networks, genetic algorithms, statistical methods, rough sets, decision trees, and fuzzy logic.

Many methods [9][70][81][218][309] have been proposed for rule extraction from neural networks. These rule extraction methods can be characterized by:

1. forms of allowed variables: continuous, discrete, or both continuous and discrete variables
   Data may have discrete, continuous, or mixed attributes. There are a few methods dealing with discrete variables [269][287][317] and continuous variables [284][305]. And, some methods deal with both continuous and discrete variables [28][102].
2. forms of extracted rule decision boundaries: hyper-rectangular, hyper-plane, hyper-ellipse, and fuzzy boundaries (Fig. 7.1).
   Crisp rule extraction methods extract rules with hyper-plane decision boundaries [111][145] or hyper-rectangular [28][154][200] decision boundaries . Rules with hyper-ellipse decision boundaries can be obtained from RBF-based rule extraction methods directly; however, the complexity of extracted rules makes this type of rule unpopular. For some applications, fuzzy rules are preferred over crisp rules in cases when approximate reasoning is desirable rather than exact reasoning. A detailed survey of fuzzy rule extraction based on neural networks can be found in [218].
3. approaches for searching rules: pedagogical, decompositional, and eclectic approaches [269].
   The pedagogical algorithms consider a learning model (such as a neural network model) as a black box and use only the activation value of input and output units in the model when extracting rules from the trained model. In contrast, the decompositional algorithms consider each unit in a learning model and unify them into the rules corresponding to the model. Compared with the former algorithms, the latter ones can utilize each single unit of trained models, and can obtain detailed rules [273]. The eclectic approach [9] is a combination of the above two categories (Fig. 7.2).
4. Learning models constructed before extracting rules: classification models and regression models.
   Most rule extraction methods are developed based on constructed classifiers, and extracted rules are used for representing decisions of the classifiers in classification and prediction. Some methods [276][288] extract rules from regression models (Fig. 7.3).

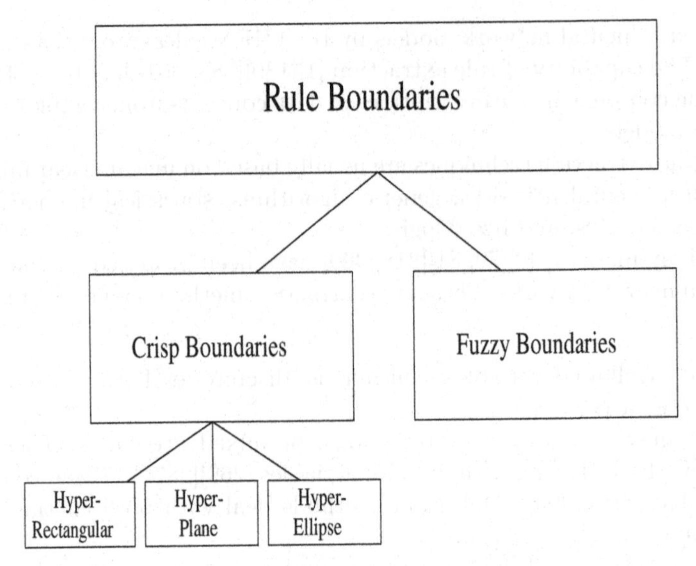

Fig. 7.1. Rule extraction boundaries.

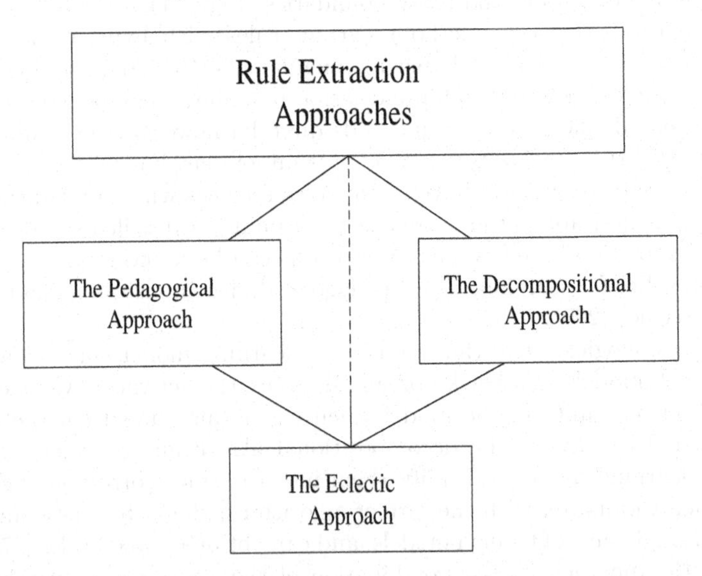

Fig. 7.2. Rule extraction approaches.

7.2 Rule Extraction Based on Classification Models

Neural networks, support vector machines (SVMs), and decision trees are popular due to their good generalization capabilities. In this section, major rule extraction techniques are reviewed according to types of learning models from which rules are extracted.

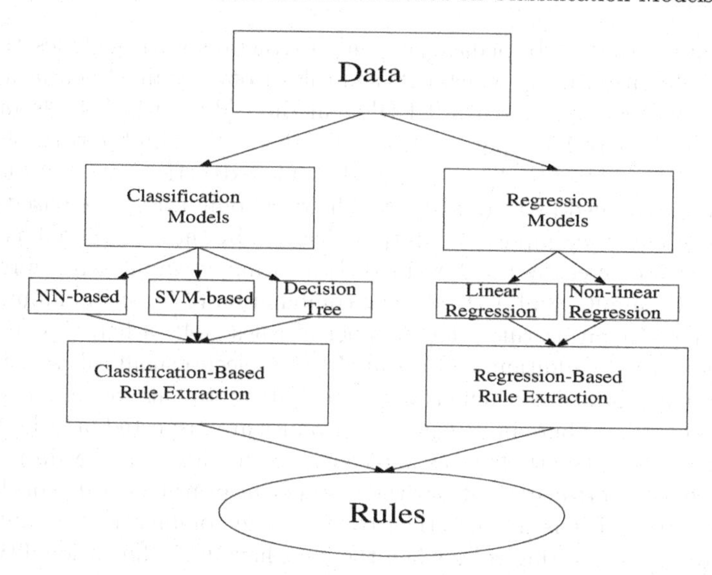

Fig. 7.3. Rule extraction based on different learning models.

7.2.1 Rule Extraction Based on Neural Network Classifiers

In classification and prediction tasks, it has been shown that neural network classifiers have remarkable generalization capabilities. Consider a data set $(\mathbf{x}_i, y_i)$, where $i = 1, 2, ..., N$, input $\mathbf{x}_i \in R^L$, and class label $y_i \in R$. Without any prior knowledge about the relationship between input samples and their corresponding class labels, neural networks are used to map the relationship to network parameters. A neural network may have multiple layers. The weights connecting one layer to the next layer are real valued. The activation functions of hidden layer neurons are non-linear. These parameters lead to complex decision functions of neural networks, which might be non-linear and non-monotonic. These factors impede explicit description of how and why an unseen sample is classified to a certain class.

Much research work has been carried out on extracting rules from MLPs [155][200][270][284]. Generally speaking, approaches to extract rules from the MLPs can be categorized according to the ways in dealing with individual neurons. In [116] and [305], rules are extracted by interpreting outputs in terms of inputs. However, hidden neurons are assumed to work independently in [116] and [305], which limits the capability of the proposed rule extraction methods. These two methods could be considered as the earliest pedagogical rule extraction approaches. The *validity interval analysis* (VIA) approach [305] improves generalization compared to Gallant's approach. *Validity intervals* are tuned as constraining inputs and outputs by detecting and excluding activation values which are not consistent with the trained neural network. This method is also considered as a pedagogical method due to the direct mapping from inputs to outputs. Other early pedagogical rule extraction work includes the RULENEG algorithm [244], which extracts rules from a trained neural network by stepwise negation, and the DEDEC method [309], which finds minimal information separating a pattern from the others based on the trained neural network. Narazaki *et al.* [226] proposed a rule extraction method analyzing the function learned by a trained NN. The rule boundaries were based on the relationship between inputs and outputs learned by the NN, as well as on the class label predicted by the NN. In [155], continuous inputs are represented by linguistic inputs, and each possible combination of linguistic inputs is examined for generating rules. The number of rules is 4^L, where L is the data dimensionality. In addition, a GA is applied to select a small rule set. Jiang *et al.* [162] transformed continuous data attributes into categorical values, and the roles of an attribute in its categorical values are inspected in order to generate rules with the attribute as antecedents. The drawback lies in that it is difficult to categorize attributes without sufficient prior knowledge of the data distribution, and it is not practical to check the combination of categorical attributes for generating rules when the data have high dimensionality.

Pedagogical rule extraction approaches have also been developed based on recurrent networks. Vahed and Omlin [323] used a symbolic learning algorithm with polynomial time to extract rules solely based on changes in inputs and outputs of a trained network. The clustering phase is eliminated in this rule extraction approach, which increases the fidelity of the extracted knowledge.

Craven and Shavlik developed the TREPAN algorithm [66] to extract rules by forming decision trees and querying the class labels of samples through trained neural networks, which can deal with both continuous and discrete attributes. Schmitz *et al.* [276] proposed an artificial neural network decision-tree algorithm (ANN-DT), in which a univariate decision tree is generated from a trained neural network.

In [162] and [361], rules are extracted from a neural network ensemble by pedagogical approaches. Similar to the method in [66], Zhou and Jiang [361] combined C4.5 with neural network ensembles to extract rules. Neural network ensembles are expected to deliver better generalization compared to single networks.

Decompositional rule extraction approaches are local methods since the basic components of neural networks, including interconnected weights and neurons, are decomposed to represent the relationship between the input and the output.

In a number of decompositional rule extraction methods [146][200], clustering techniques are used for grouping activation values of hidden neurons. In these approaches, the connected weights and activation values are approximated according to clustering results.

GAs have been widely used for practical problem solving and for scientific modelling. With the capability of searching for desirable solutions in the problem space, GAs have been employed for extracting rules from neural networks. Fukumi and Akamatsu [111] used GAs to prune the connections in neural networks before extracting rules. Hruschka and Ebecken [146] proposed a clustering genetic algorithm (CGA) to cluster the activation values of the hidden units of a trained neural network. Rules were then extracted based on the results from the CGA. Ishibuchi *et al.* [152][153][154][155][156] used GAs to obtain concise rules by selecting important members from the rules extracted from a neural network.

Decision trees are often combined with neural networks in both pedagogical and decompositional rule extraction approaches [273][283]. In the decompositional approach proposed in [273], neural networks are first trained to extract the essential relationship between the input and the output. The relationship is thus embedded in interconnected weights and hidden neurons of trained neural networks. Then, decision trees are applied to decompose the relationship between inputs and hidden neurons, as well as the relationship between hidden neurons and outputs. The results from decision trees are combined to deliver rules.

7.2.2 Rule Extraction Based on Support Vector Machine Classifiers

In recent years, support vector machines (SVMs) [39][40][42][163] have attracted lots of interest for their capability of solving classification and regression problems. Successful applications of SVMs have been reported in various areas, such as communication [122], time-series prediction [119], and bioinformatics [34][222]. In many applications, it is desirable to know not only the classification decisions but also what leads to the decisions. However, SVMs offer little insight into the reasons why SVMs offer their final results.

In [231], rules are extracted after clustering. Distances from support vectors to a cluster center are checked to generate rules based on the cluster. The RulExSVM method [108] extracts hyper-rectangular rules which are then fine tuned, with redundant rules merged to produce a compact rule set. This will be described in a later chapter.

7.2.3 Rule Extraction Based on Decision Trees

As stated in previous paragraphs, decision trees are employed for facilitating rule extraction from neural networks. Rules can be obtained from decision tree classifiers directly since each distinct path through the decision tree produces a distinct rule. In order to generate rules, each path is traced in the decision tree, from the root node to the leaf node, recording the outcome as the antecedents and the leaf-node classification as the consequences. Decision trees are easy to construct automatically from labelled instances. Two well-known programs for constructing decision trees are C4.5 [249] and CART (classification and regression tree) [30]. Decision trees can be regarded as rule-based systems.

Knowledge learned by decision trees can be transformed into expressive rules easily. However, decision trees are prone to errors when data are noisy.

Decision trees [126][321] can form concise rules, in contrast to neural networks. However, the accuracy of decision trees is often lower than neural networks for noisy data and it is difficult for decision trees to tackle dynamic data. Zhao [360] constructed a decision tree with each node being an expert neural network, therefore combining the advantages of both the decision tree and the neural network. Tsang *et al.* [315] and Umano *et al.* [321] combined neural networks with decision trees to obtain better performance in rule extraction.

7.2.4 Rule Extraction Based on Regression Models

Most rule extraction methods for analyzing data or explaining functions learned by trained models are developed based on classification models. Classification models have discrete outputs, i.e., the categories of the inputs. Regression and approximation models are constructed for approximating continuous outputs. There are fewer methods developed for extracting rules from learning models with continuous outputs.

Setiono *et al.* [288] proposed a method REFANN (rule extraction from function approximating neural networks) to extract rules from trained neural networks for non-linear function approximation or regression. The trained neural network is first pruned by removing redundant inputs and hidden neurons. Then, either a three-piece or a five-piece linear function is used to approximate the continuous activation function of each hidden neuron. Finally, the input space is divided into subregions. In each subregion, the function values of samples are computed as a linear function of the inputs. That is, the antecedents of a rule are a subregion represented by attributes, and the consequence of the rule is a linear function of inputs as the final approximation of the non-linear function learned by the neural network.

Tsukimoto [318] extracted rules from regression models and mathematical formulae. In the rule extraction technique called LRA (logical regression analysis) [318], the trained neural network is decomposed into neurons. In this algorithm, the output of each neuron in a trained neural network is assumed to be monotone increasing in the algorithm [318].

7.3 Components of Rule Extraction Systems

Generally, a rule extraction system includes the following components:

1. *Data collection*
   Data are accumulated in diverse organizations. For example, large volumes of data are produced in transactions on the Internet, in supermarkets, and in banks. Valuable information is hidden in huge volumes of data, which

calls for intelligent and efficient techniques for discovering knowledge in order to make better decisions, improve profits, save resources, and reduce labor costs.

2. *Data preprocessing*
   Diverse data formats and data objects are stored in data repositories. Many variables (attributes) are collected for the purpose of illustrating the concept of objects. However, not all attributes are necessary for analyzing data, i.e., some irrelevant or unimportant data may be included in data sets. In order to remove irrelevant information which may interfere with the data analysis process, dimensionality reduction is widely explored for both memory constraint and speed limitation. Hence, the following preprocessing of the data is needed.
   - *Feature selection*: much research work [27][168][198] has been carried out in choosing a feature subset to represent the concept of data with the removal of irrelevant and redundant features.
   - *Normalization*: for a neural network, input values are usually normalized to lie in [0, 1]. Nominal inputs should be transformed into numerical ones.
3. *The selection of rule extraction tools*
   Decision trees, neural networks, genetic algorithms, and fuzzy logic are powerful data mining tools. Rules can be extracted from these tools.
4. *The expression of the extracted rules*
   Usually, the rules are in IF–THEN forms. The premise parts of the rules are composed of different combinations of inputs. There are three kinds of rule decision boundaries:
   - *hyper-rectangular,*
   - *hyper-plane,*
   - *hyper-ellipse.*

   They are shown in Figs. 7.4–7.6. The hyper-rectangular boundary is the simplest. However, since the distributions of data may be different for different problems, different decision boundaries or combinations of different boundaries may be required (see Fig. 7.7) for different problems. Finding the most efficient decision boundary type will be one of our future tasks.

7.4 Rule Extraction Combining GAs and the RBF Neural Network

We describe a novel decompositional method which we proposed earlier [100] to extract rules by combining genetic algorithms (GAs) and radial basis function (RBF) neural networks.

Firstly, the data are classified by an RBF classifier. Each hidden neuron of the RBF classifier corresponds to a rule. During training of the RBF network, we allow for large overlaps between clusters of the same class to

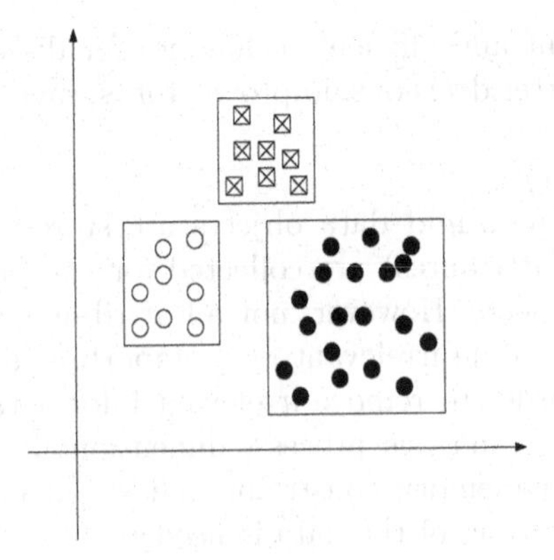

Fig. 7.4. Hyper-rectangular decision boundaries.

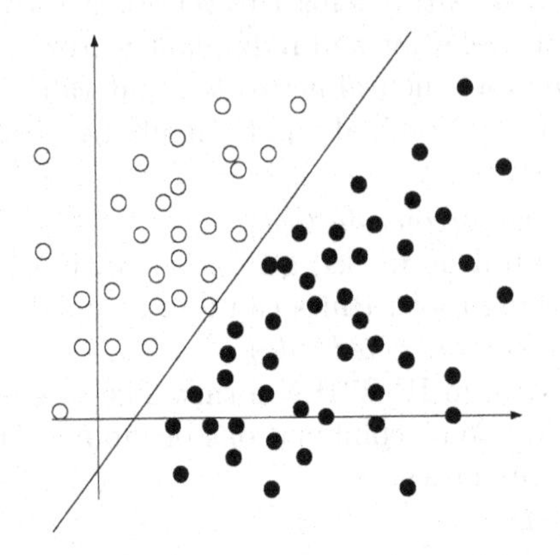

Fig. 7.5. Hyper-plane decision boundaries.

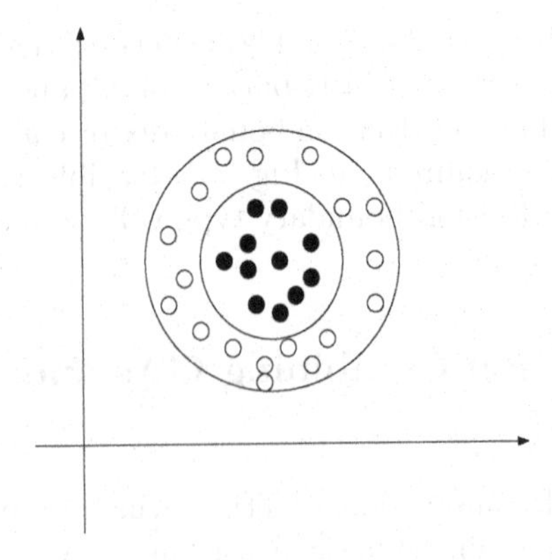

Fig. 7.6. Hyper-ellipse decision boundaries.

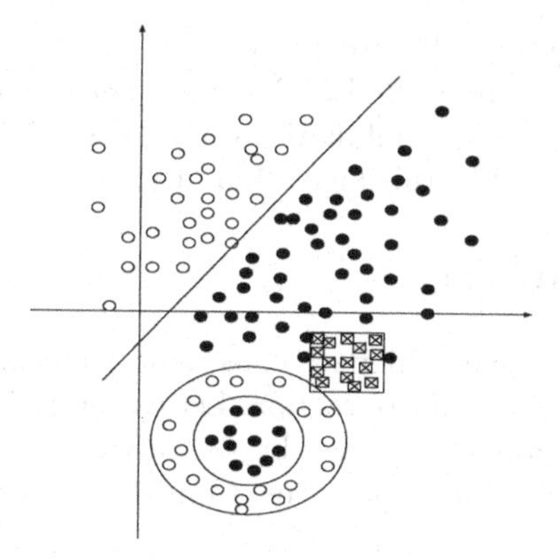

Fig. 7.7. Decision boundaries of mixed types.

reduce the number of hidden neurons while maintaining the classification accuracy. The weights connecting the hidden neurons with the output neurons are then pruned in order to clearly show which output neuron each hidden neuron mainly serves. Next, centers of the kernel functions are used as initial conditions when searching for the premises of rules with GAs. The interval for each attribute in the condition part of each rule is encoded into a GA chromosome. The fitness of a chromosome is determined by the accuracy of the extracted rules. The rule set obtained is further processed to remove redundant information. Our method leads to rules with hyper-rectangular decision boundaries directly without the need for an intermediate step to transform continuous attributes into discrete ones, unlike some existing methods based on the multi-layer perceptron (MLP) [28][145].

7.4.1 The Procedure of Rule Extraction

Our GA-based rule extraction system consists of the following components:

1. *Data collection*
   Data sets from the UCI database are used to demonstrate algorithms.
2. *Data preprocessing*
   For RBF neural networks, input values are usually normalized to lie in [0, 1]. Hence, each attribute of the data is normalized to lie in [0, 1]. If there are nominal attributes, they are transformed into numerical ones.
3. *The selection of rule extraction tools*
   In our algorithm, RBF neural networks and genetic algorithms are combined for rule extraction. After preprocessing the data, the normalized data are input to an RBF neural network. The RBF neural network is trained as stated in Chap. 4. The weights connecting the hidden layer and the output layer are then pruned. Each hidden unit corresponds to a rule. The condition part of each rule is determined by GA searching.

4. *The expression of the extracted rules*
   Our extracted rules are in IF–THEN forms. The premise parts of the rules are composed of various combinations of inputs. The decision boundaries of the rules extracted in this algorithm are hyper-rectangular.

7.4.2 Simplifying Weights

The weights connecting the hidden neurons (units) and the output neurons are important information for rule extraction. Let us determine the output neuron which a hidden neuron mainly serves by simplifying the weights between the hidden neurons and the output neurons. Consider the weight matrix:

$$W = \begin{pmatrix} w_{11} & w_{12} & \cdots & w_{1M} \\ w_{21} & w_{22} & \cdots & w_{2M} \\ \cdots & \cdots & \cdots & \cdots \\ w_{m1} & w_{m2} & \cdots & w_{mM} \end{pmatrix} , \tag{7.1}$$

where w_{ik} ($i = 1, ..., m$, $k = 1, ..., M$) is the weight which connects hidden neuron i with output neuron k. m is the number of hidden neurons and M is the number of class labels in the RBF neural network. We simplify the matrix W to:

$$W_1 = \begin{pmatrix} 0 \cdots & w_{1k_1} & \cdots & 0 \\ 0 \cdots & w_{2k_2} & \cdots & 0 \\ \cdots & \cdots & \cdots & \cdots \\ 0 \cdots & \cdots & w_{mk_M} & 0 \end{pmatrix} , \tag{7.2}$$

where w_{ik_i} is the maximum value of row i ($i = 1, ..., M$, k_i corresponds to the index of the maximum value of row i). Thus, W_1 reflects the corresponding output neuron which a hidden neuron mainly serves. Simulations show that one weight in each row is significantly larger than the other weights in this row.

7.4.3 Encoding Rule Premises Using GAs

The rules extracted here are in an IF–THEN form. Rule i (corresponding to hidden neuron i) is written as follows:

IF attribute 1 is within the interval (L_{1i}, U_{1i})
AND attribute 2 is within the interval (L_{2i}, U_{2i})
.
.
.
AND attribute n is within the interval (L_{ni}, U_{ni})
THEN the class label of the input pattern is k_i.

Here U_{ji} and L_{ji} are the upper limit and the lower limit of interval j in rule i, respectively. The center of the ith hidden unit is $\overrightarrow{\mu_i} = \{\mu_{1i}, \mu_{2i}, ..., \mu_{ji}, ..., \mu_{ni}\}$.

U_{ji} and L_{ji} are set according to the trained RBF classifier. Initially, U_{ji} is randomly generated according to the uniform distribution within the range $[\mu_{ji}, 1]$, and L_{ji} is randomly generated according to the uniform distribution within the range $[0, \mu_{ji}]$.

The input data space is separated into several subspaces through training the RBF neural network. Each subspace is represented by a hidden neuron of the RBF neural network and is a hyper-ellipse. Since the decision boundary of our rules is hyper-rectangular, we use a GA to search for the premise parts of the rules. There are as many rules as hidden neurons. Hence, an efficient architecture of the RBF neural network with fewer hidden units leads to a compact set of rules.

We encode real value p_{ji} $(p = U, L)$ using k binary bits:

$$G^{(p)}{}_{ji} = \{g_k, g_{k-1}, ..., g_i, ..., g_2, g_1\}, \ g_i = 0, 1, \ i = 1, 2, ..., k. \tag{7.3}$$

The relationship between p_{ji} and $G^{(p)}{}_{ji}$ is as follows:

$$p_{ji} = B^{(p)}{}_{ji}/(2^k - 1), \tag{7.4}$$

where $B^{(p)}{}_{ji}$ is the decimal value corresponding to $G^{(p)}{}_{ji}$:

$$B^{(p)}{}_{ji} = g_k * 2^{k-1} + g_{k-1} * 2^{k-2} + ... + g_2 * 2^1 + g_1 * 2^0. \tag{7.5}$$

A chromosome in the population pool can be represented as a one-dimensional binary string:

$$(G^{(U)}{}_{11}, G^{(L)}{}_{11}, ..., G^{(U)}{}_{n1}, G^{(L)}{}_{n1}, ..., G^{(U)}{}_{1m}, G^{(L)}{}_{1m}, ..., G^{(U)}{}_{nm}, G^{(L)}{}_{nm}). \tag{7.6}$$

7.4.4 Crossover and Mutation

We use the roulette wheel selection to select chromosomes in each generation, i.e., the selection probability is proportional to each chromosome's fitness. The two-point crossover is used in our algorithm. Two points are randomly located in each of the two parents. The two parts of the parent chromosomes between the two pairs of points are then exchanged to generate new offspring. The probability of crossover is around 80%.

Mutation can prevent fixation at some particular locus. A locus in the parent chromosome is selected randomly and the bit at the position is flipped, i.e., if the original bit is 0, it is replaced by 1, and vice versa. Usually, the mutation rate is relatively small to avoid random variations. However, at later generations, the number of identical members increases, which leads to a stagnant state. In order to break stagnant states for searching for optimal results, we use a dynamic mutation rate, i.e., if the number of identical members in a population exceeds a certain percentage, the mutation rate is increased by a certain amount. In [195], a dynamic mutation rate was introduced by maintaining that the sum of the mutation rate and the crossover rate is 1, i.e., if the mutation rate increases, the crossover rate decreases. In addition, 'elitism' is used to retain the best members in the population pool.

7.4.5 Fitness Function

Our fitness function is:

$$F(\{(L_{ji}, U_{ji})\}) = 1 - E(\{(L_{ji}, U_{ji})\}), \quad (7.7)$$

where $E(\{(L_{ji}, U_{ji})\})$ is the error rate of the extracted rules and

$$\{(L_{ji}, U_{ji})\} \equiv \{(L_{11}, U_{11}), (L_{12}, U_{12}), ..., (L_{nm}, U_{nm})\}. \quad (7.8)$$

Each chromosome in the population pool corresponds to a rule set. The accuracy of the rule set is calculated to evaluate the fitness level of each chromosome. The better the fitness of the chromosome, the lower the error rate of its corresponding rule set.

7.4.6 More Compact Rules

We have so far implicitly assumed that all attributes contribute to each rule. However, some attributes contribute little to the description of the data, or even do not contribute to the description at all, i.e., those attributes are irrelevant or redundant for some rules and hence should be removed.

Based on our methods in the preceding sections, if there are m hidden neurons in the trained RBF neural network, m rules will be adjusted by a GA. Originally, the premise part of each rule is composed of n conditions (n is the number of attributes). For example, for rule i, condition j is in this form: 'IF attribute j is within the interval (L_{ji}, U_{ji})'. If an attribute does not affect a certain rule, its variation in the rule will not affect the final rule decision. It is desirable to remove this attribute from the rule. Assume that the minimum value and the maximum value in the data for attribute j are $\min_j$ and $\max_j$, respectively. If the following conditions are satisfied:

$$L_{ji} \leq \min_j, \quad (7.9)$$

and

$$U_{ji} \geq \max_j, \quad (7.10)$$

condition j will be removed from the original rule i. Thus, rule i will be replaced by a new rule with fewer conditions.

7.4.7 Experimental Results

We use the Iris and Thyroid data sets [223] to demonstrate our method. Each data set is divided into three parts, i.e., training, validation, and test sets. 90 patterns of the Iris data set are used for training, 30 patterns for validation, and 30 patterns for testing. There are 215 patterns in the Thyroid data set. 115 patterns are used for training, 50 patterns for validation, and 50 patterns for testing.

We set the ratio between the number of in-class patterns and the total patterns in a cluster as $\theta = 100\%$ (the initial θ-criterion) in our experiments.

Iris Data Set

Different population sizes may affect the rule extraction based on GAs. The results shown in Table 7.1 are the average values of five independent experiments with randomly selected initial cluster centers. The smallest number of hidden neurons required to construct an RBF neural network classifier is three for the Iris data set.

Next we use GAs to search for optimal rules based on the compact RBF neural network obtained above. The results in Table 7.1 show that the number of generations needed for reaching the top rule accuracy is reduced with the increase of population size.

Table 7.1. The results under different population sizes for the Iris data set (average of five runs).

Population size	40	80	160	200	240
Rule error rate	6.67%	6.67%	3.3%	2.67%	2.67%
Number of generations needed to reach the accuracy	33.5	31	13.2	16	9

Thyroid Data Set

In Chap. 4, it had shown that when large overlaps among clusters of the same class are permitted, both the number of hidden neurons and the classification error rate are reduced. For the Thyroid data set, at least six hidden neurons are needed.

Referring to the weight matrix of the Thyroid data set shown in Chap. 6, the simplified weight matrix is:

$$W_1 = \begin{pmatrix} 0.91 & 0.00 & 0.00 \\ 0.00 & 1.74 & 0.00 \\ 0.97 & 0.00 & 0.00 \\ 0.00 & 0.00 & 1.61 \\ 0.00 & 0.87 & 0.00 \\ 0.00 & 0.93 & 0.00 \end{pmatrix} .$$

Next we use a GA to search for optimal rules based on the compact RBF neural network obtained above. The results with different population sizes are compared in Table 7.2. It shows that the number of generations needed for approaching the optimal results decreases and the rule accuracy increases, but eventually saturates, as the size of population increases. Chromosomes in the initial population are initialized to be around the centers of the Gaussian kernel functions. The number of bits for each ${G^{(p)}}_{ji}$ in Eq. (7.3) is $k = 6$. The crossover probability is 90% and the mutation rate is dynamic. The initial mutation rate is 10%. If the number of identical members in a population exceeds 25%, the mutation rate is increased by 1%. The number of elite chromosomes, which remain unchanged and live from one generation to the next, is two. The GA searching procedure is stopped if the accuracy of the extracted rules does not change by more than 0.5% for the validation set for five consecutive generations.

Table 7.2. The results under different population sizes for the Thyroid data set (average of five runs).

Population size	40	80	160	200	240	280
Rule error rate	25.2%	24.3%	22%	21%	20%	20%
Number of generations needed to reach the accuracy	36.5	28	23	38	38.2	26

After the searching procedure of the GA based on the smallest trained RBF neural network, the redundant premises in each rule will be checked according to Eqs. (7.9) and (7.10). We obtain three symbolic rules for the Iris data set. The average number of premises in each rule is two. The accuracy of the symbolic rules that we obtain through the proposed method is 97.33% for the Iris data set. There are six rules for the Thyroid data set. The average number of premises in each rule is 4.2, and the accuracy of the extracted rules is 80%. In addition, there is a default rule for each data set. The dominant class will be the class label for the default rule if the data set is biased. If the classes in the data set have the same number of patterns, then the default rule corresponds to the last class. The smallest number of generations needed to obtain the best accuracy is 200 for the Iris data set and 240 for the Thyroid data set.

The linguistic rules for describing the two data sets obtained are as follows. For the Iris data set, three rules are obtained using our method; the accuracy is 97.33% for the test data set. These rules are not the same as rules extracted from the MLP [28][145], but have the same accuracy.

Rule 1:

IF the petal length is within the interval (0.00, 2.08)
AND the petal width is within the interval (0.00, 1.11)

THEN the class label is Setosa.

Rule 2:

IF the petal length is within the interval (2.08, 5.26)
AND the petal width is within the interval (0, 1.71)
THEN the class label is Versicolor.

Rule 3:

IF the petal length is within the interval (0.77, 6.9)
AND the petal width is within the interval (1.79, 2.5)
THEN the class label is Virginica.

Default rule:

the class label is Virginica.

For the Thyroid data set, six rules are obtained, and the accuracy is 80% for the test data set.

Rule 1:

IF attribute 1 is within the interval (116.64, 118.94)
AND attribute 3 is within the interval (0.00, 7.63)
AND attribute 4 is within the interval (0.00, 56.40)
AND attribute 5 is within the interval (0.00, 35.36)
THEN the class label is normal.

Rule 2:

IF attribute 1 is within the interval (33.22, 144.00)
AND attribute 2 is within the interval (13.54, 25.30)
AND attribute 3 is within the interval (0.00, 9.03)
AND attribute 4 is within the interval (0.00, 30.70)
THEN the class label is hyper.

Rule 3:

IF attribute 1 is within the interval (63.55, 140.36)
AND attribute 2 is within the interval (0.00, 13.35)
AND attribute 3 is within the interval (0.00, 7.20)
AND attribute 4 is within the interval (8.42, 8.93)
AND attribute 5 is within the interval (0.00, 53.84)
THEN the class label is normal.

Rule 4:

IF attribute 1 is within the interval (68.40, 144.00)
AND attribute 2 is within the interval (0.00, 19.06)
AND attribute 4 is within the interval (0.15, 15.94)
THEN the class label is hypo.

Rule 5:

IF attribute 1 is within the interval (11.19, 144.00)
AND attribute 2 is within the interval (24.11, 24.11)
AND attribute 4 is within the interval (0.00, 42.15)
AND attribute 5 is within the interval (0.00, 11.37)
THEN the class label is hyper.

Rule 6:

IF attribute 1 is within the interval (51.26, 108.86)

AND attribute 2 is within the interval (14.57, 25.30)
AND attribute 3 is within the interval (7.33, 10.00)
AND attribute 4 is within the interval (0.00, 34.74)
AND attribute 5 is within the interval (0.00, 18.65)
THEN the class label is hyper.

Default rule:

the class label is normal.

Halgamuge *et al.* [128] extracted rules based on RBF neural networks; however, five or six rules were needed to represent the concept of the Iris data set (the accuracy is not available). Huber and Berthold [150] used eight rules to represent the Iris data set (the accuracy is not available). In order to obtain a small rule base, unimportant rules were pruned according to ranking [150]; however, the accuracy of the rules was reduced at the same time. McGarry *et al.* [212][213][214] extracted rules from RBF neural networks directly based on the parameters of Gaussian kernel functions and weights. In [212], the accuracy reached 100%, but the number of rules was large (for the Iris data set, 53 rules are needed). In [213] and [214], the number of rules for the Iris data set was small, i.e., three, but the accuracy of the extracted rules was only 40% and around 80%, respectively. The rule set for the Iris data set extracted by McGarry*et al.* [213] is included in Appendix A for comparison. The results of the extracted rules for the Thyroid data set using other methods are not available.

In order to evaluate the complexity of the extracted rule set, a complexity measure [115] was calculated for comparison. The complexity measure C was defined as:

$$C = 0.6NR + 0.4NP, \tag{7.11}$$

where NR is the number of rules and NP is the number of premises (Table 7.3).

Much work has been carried out on extracting rules using MLPs [28][145]. However, prior knowledge about how to divide the range of each attribute into several parts is needed. In most cases, it is difficult to obtain this knowledge.

In addition, in [28], no explicit rule corresponds to the Virginica class, i.e., the patterns belonging to the Virginica class cannot be selected from the data set independently, in contrast to other methods in Table 7.3. The rules extracted by Bologna *et al.* [28] are shown in Appendix A. The rules extracted by Hruschka and Ebecken [145] are also shown in Appendix A. In the presented algorithm, rules with hyper-rectangular decision boundaries are obtained directly without the need for transforming continuous attributes into discrete ones.

7.4.8 Summary

We have described a novel rule-extraction algorithm that we proposed earlier by combining GAs and RBF networks. Rule extraction is carried out from

Table 7.3. A comparison of results for the Iris data set obtained with different methods.

Methodology	Rule accuracy	Complexity [115]	Type of decision boundary
Modified RX algorithm based on MLP [145]	97.33%	3.4	Hyper-plane
Inputs are transformed into discrete ones artificially based on IMLP [28]	97.33%	2.2	Hyper-rectangular
Based on RBF [214]	80%	3.4	Hyper-rectangular
Based on RBF [212]	100%	32.2	Hyper-rectangular
Our algorithm combining GA and RBF	97.33%	2.6	Hyper-rectangular

a compact RBF classifier in order to explain and represent the concept of data in a concise way. First, a compact RBF network is obtained by allowing for large overlaps among the clusters belonging to the same class. Next, the weights between the hidden layer and the output layer are simplified. Then, the interval for each input in the condition part of a rule is determined by a GA. Experimental results show that our rule extraction technique is simple to implement, and concise rules with high accuracy are obtained. In addition, rules extracted by our algorithm have hyper-rectangular decision boundaries, which are desirable due to their explicit perceptibility.

7.5 Rule Extraction by Gradient Descent

7.5.1 The Method

The objective of tuning the rule premises is to determine the boundaries of rules so that a high rule accuracy is obtained for the test data set. In this section, we describe an algorithm to extract rules from trained RBF neural networks using the gradient descent method, which we proposed earlier [105].

Before starting the tuning process, all of the premises of the rules must be initialized. Assume that the number of attributes is n. The number of rules equals the number of hidden neurons in the trained RBF network. The number of the premises of the rules equals n. The upper limit U_{ji} and the lower limit L_{ji} of the jth premise in the ith rule are initialized according to the trained RBF classifier as:

$$U_{ji}^{(0)} = \mu_{ji} + \sigma_i, \tag{7.12}$$

$$L_{ji}^{(0)} = \mu_{ji} - \sigma_i, \tag{7.13}$$

where μ_{ji} is the jth coordinate of the center of the ith kernel function. σ_i is the width of the ith kernel function.

We introduce the following notation. Suppose that $\eta^{(t)}$ is the tuning rate at time t. Initially $\eta^{(0)} = 1/N_I$, where N_I is the number of iteration steps for adjusting a premise. N_I is set to be 20 in our experiments, i.e., the smallest changing scale in one tuning step is 0.05, which is determined empirically. E is the rule error rate. Denote

$$Q_{ji}^{(t)} \equiv \frac{\partial E}{\partial U_{ji}} |_t, \tag{7.14}$$

$$A_{ji}^{(t)} \equiv \frac{\partial E}{\partial L_{ji}} |_t . \tag{7.15}$$

$U_{ji}^{(t)}$ and $L_{ji}^{(t)}$, the upper and lower limits at time t, are tuned as follows.

$$U_{ji}^{(t+1)} = U_{ji}^{(t)} + \Delta U_{ji}^{(t)}, \tag{7.16}$$

$$L_{ji}^{(t+1)} = L_{ji}^{(t)} + \Delta L_{ji}^{(t)}. \tag{7.17}$$

Initially, we let

$$\Delta U_{ji}^{(0)} = \eta^{(0)}, \tag{7.18}$$

$$\Delta L_{ji}^{(0)} = -\eta^{(0)}. \tag{7.19}$$

Subsequent $\Delta U_{ji}^{(t)}$ and $\Delta L_{ji}^{(t)}$ are calculated as follows.

$$\Delta W_{ji}^{(t)} = \begin{cases} \eta^{(t)} & \text{if } Q_{ji}^{(t-1)} < 0, \\ -\eta^{(t)} & \text{if } Q_{ji}^{(t-1)} > 0, \\ \Delta W_{ji}^{(t-1)} & \text{if } Q_{ji}^{(t-1)} = 0, \\ -\Delta W_{ji}^{(t-1)} & \text{if } Q_{ji}^{(t-1)} = 0 \text{ for} \\ & \frac{1}{3}N_I \text{ consecutive iterations,} \end{cases} \tag{7.20}$$

where $W = U, L$. When $Q_{ji}^{(t)} = 0$ consecutively for $\frac{1}{3}N_I$ time steps, which means that the current direction of premise adjustment is fruitless, $\Delta W_{ji}^{(t)}$ changes its sign as shown in the fourth line of Eq. (7.20). In this situation, we also let $\eta^{(t)} = 1.1\eta^{(t-1)}$, which helps to keep the progress from being trapped. Otherwise, $\eta^{(t)}$ remains unchanged.

Compared with the technique proposed by McGarry *et al.* [212][213][214], a higher accuracy with concise rules is obtained with this method. In [212][214], the input intervals in rules are expressed by the following equations:

$$X_{upper} = \mu_i + \sigma_i - S, \tag{7.21}$$

$$X_{lower} = \mu_i - \sigma_i + S. \tag{7.22}$$

Here X_{upper} is the upper limit of the premise of a rule, and X_{lower} is the lower limit. S is the feature 'steepness', which was discovered empirically to be about 0.6 by McGarry *et al.*. μ_i is the n-dimensional center location of rule i and σ_i is the width of the receptive field. We note that the empirical parameter S may vary from data sets to data sets.

Two rule-tuning stages are used in our method. In the first tuning stage, the premises of m rules (m is the number of hidden neurons of the trained RBF network) are adjusted using gradient descent to minimize the rule error rate. Some rules do not contribute to the improvement of rule accuracy, which is due to the following reason. The input data space is separated into several subspaces through training the RBF neural network. Each subspace is represented by a hidden neuron of the RBF neural network and is a hyper-ellipse. The decision boundary of our rules is hyper-rectangular. We use gradient descent for searching the premise parts of rules. Since overlaps (Fig. 7.8(a)) exist between clusters of the same class, some hidden neurons may be overlapped completely when a hyper-rectangular rule is formed using gradient descent (see Fig. 7.8(b)). Thus, the rules overlapped completely are redundant for representing data and should be removed from the rule set. It is expected that this action will not reduce the rule accuracy. The number of rules should be fewer than the number of hidden neurons.

Based on the results of the first tuning stage, the second tuning stage removes irrelevant and unimportant features by calculating an importance factor, i.e., variations of rule accuracy on validation data when tuning the feature. We set the importance factor threshold for removing a feature as 1%, i.e., if the rule accuracy of the validation set does not decrease by 1% when tuning a feature on the first training stage, the feature will be considered to be unimportant, and will be deleted from the data set. Rules with boundaries completely overlapped by other rules are redundant and will be removed.

7.5.2 Experimental Results

The Thyroid, Breast cancer, and Glass data sets available at the UCI database [223] are used to demonstrate our method.

Table 7.4 shows that when large overlaps among clusters of the same class are permitted, both the number of hidden neurons and the classification error rate are reduced.

Thyroid Data Set

Four rules (Table 7.5) are extracted for the Thyroid data set by the method described in this section. The average number of premises in each rule is three, and the accuracy of the extracted rules is 92% for the test data set. Experimental results show that better rule accuracy is obtained by this rule extraction method compared with the GA-based rule extraction method described in Sect. 7.4. The rules for the Thyroid data set are as follows:

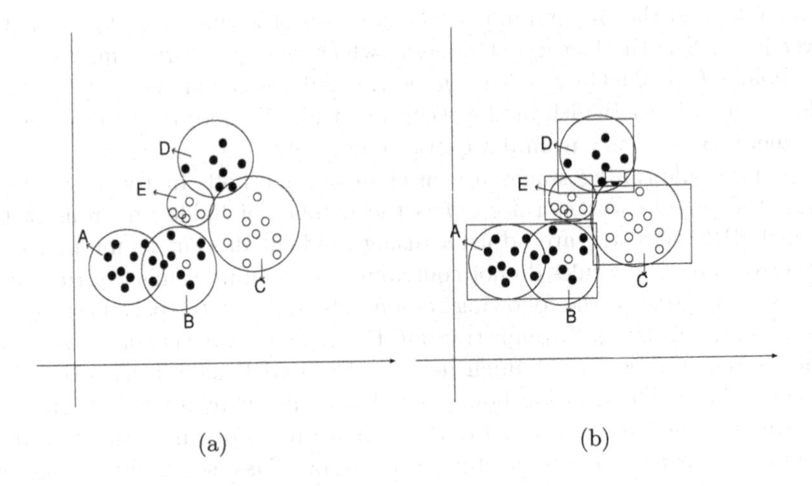

Fig. 7.8. (a) Clusters in an RBF network, (b) hyper-rectangular rule decision boundaries corresponding to the clusters.

Table 7.4. Reduction in the number of hidden units in the RBF network when large overlaps are allowed between clusters for the same class.

Results		Thyroid	Breast cancer	Glass
Classification accuracy	Small overlap	94%	97.08%	78.41%
	Large overlap	95.2%	98.54%	85.09%
Number of hidden units	Small overlap	14.4	31	13
	Large overlap	8	11	10

Rule 1:
IF attribute 2 is within the interval [11.97, 22.57]
AND attribute 3 is within the interval [2.50, 10]
AND attribute 5 is within the interval [0, 13.62]
THEN the class label is hyper-thyroid.

Rule 2:
IF attribute 2 is within the interval [15.49, 25.3]
AND attribute 3 is within the interval [1.3, 10]
AND attribute 5 is within the interval [0, 13.73]
THEN the class label is hyper-thyroid.

Rule 3:
IF attribute 2 is within the interval [0, 4.62]
AND attribute 3 is within the interval [0, 2.62]
AND attribute 5 is within the interval [0, 17.11]
THEN the class label is hypo-thyroid.

Rule 4:
IF attribute 2 is within the interval [0.26, 7.81]
AND attribute 3 is within the interval [0.0, 2.61]
AND attribute 5 is within the interval [8.78, 55.73]
THEN the class label is hypo-thyroid.
Default rule:
the class label is normal.

Table 7.5. Rule accuracy and numbers of rules for the Thyroid data set by the gradient descent method.

Results		Thyroid
Rule accuracy	Training accuracy	98.26%
	Validation accuracy	96%
	Testing accuracy	92%
The number of premises/rule		3
The number of rules		4

Glass Data Set

There are nine attributes, six classes, and 214 patterns in the Glass data set. For a comparison with the results in [126], only attributes 2, 3, 4, 5, 6, 7, and 8 were used in the Glass data set. Six rules (Table 7.6) are extracted for

Table 7.6. Rule accuracy and numbers of rules for the Glass data set by the gradient descent method.

Results		Glass
Rule accuracy	Training accuracy	84.85%
	Validation accuracy	86.21%
	Testing accuracy	86.21%
The average number of premises per rule		3.33
The number of rules		6

the Glass data set by our method. The average number of premises in each rule is 3.33, and the accuracy of the extracted rules is 86.21%. In [126], two rule extraction results are shown for the same Glass data set. A rule accuracy of 83.88% was obtained based on the C4.5 decision tree. A rule accuracy of 83.33% was obtained by the GLARE rule extraction method based on the MLP. Hence, experimental results show that better rule accuracy is obtained by our rule extraction method.

Breast Cancer Data Set

Based on our method, we obtain four symbolic rules (Table 7.7) for the Breast cancer data set. The average number of premises in each rule is two. The accuracy of the symbolic rules obtained through our method is 96.35% for the test data set. In comparison, Setiono [286] extracted 2.9 rules and obtained 94.04% accuracy for the Breast cancer data set based on the pruned MLP.

Table 7.7. Rule accuracy and numbers of rules for the Breast cancer data set by the gradient descent method.

Results		Breast cancer
Rule accuracy	Training accuracy	95.35%
	Validation accuracy	95.62%
	Testing accuracy	96.35%
The average number of premises per rule		2
The number of rules		4

7.5.3 Summary

We have described a novel rule-extraction algorithm from RBF networks based on the gradient descent method. First, a compact RBF network is obtained by allowing for large overlaps among the clusters belonging to the same class. Second, the rules are initialized according to the training result. Next, premises of each rule are tuned using gradient descent theory. The unimportant rules which do not affect the rule accuracy will be removed from the rule set. The unimportant features will also be deleted from the data set based on the results obtained in the first tuning stage. Fourth, rules left will be tuned using gradient descent theory again. Experimental results show that our rule extraction technique is simple to implement, and concise rules with high accuracy are obtained. In addition, rules extracted by our algorithm have hyper-rectangular decision boundaries, which are desirable due to their explicit perceptibility. The approach eliminates the need for an error-prone transformation from continuous attributes into discrete ones as required in MLP-based methods.

7.6 Rule Extraction After Data Dimensionality Reduction

Data dimensionality reduction (DDR) is usually carried out before inputting patterns to classifiers. In order to obtain good results in data mining, careful selections of relevant data are desirable. Irrelevant or redundant attributes interfere with knowledge discovery from data sets. In this section, we carry out rule extraction after data dimensionality reduction shown in Chap. 5. According to the attribute ranking results, the attribute subsets which lead to the best classification results are selected and used as inputs to a classifier, such as an RBF neural network used in this book. The complexity of the classifier can thus be reduced and its classification performance improved. The results are input to our rule extraction system to discover knowledge from data sets. Rules with hyper-rectangular decision boundaries are extracted based on the trained RBF neural networks and DDR using the gradient descent method.

7.6.1 Experimental Results

Iris Data Set

150 patterns of the Iris data set are divided into three sets, i.e., 90 patterns for training, 30 for validation, and 30 for testing. For the Iris data set, based on the attribute subset $\{3, 4\}$ selected in Chap. 5, two rules are obtained with two antecedents per rule. The accuracy is 100% for the test data set (Table 7.8). We compare our rule extraction results for the Iris data set with other methods in Table 7.9.

Table 7.8. Rule accuracy and numbers of rules for the Iris data set based on DDR.

Rule accuracy	Training accuracy	100%
	Validation accuracy	96.67%
	Testing accuracy	100%
The number of premises/rule		2
The number of rules		2

Monk3 Data Set

There are six attributes in the Monk3 data set. The Monk3 data set has a training set with 122 patterns and a test set with 421 patterns. We divide the test set into 200 patterns for validation and 221 patterns for testing.

For the Monk3 data set, based on the attribute subset $\{2, 4, 5\}$ selected in Chap. 5, we obtain three rules with three antecedents per rule (Table 7.10). The rule accuracy is 98% for the test data set. Setiono [286] extracted two rules, 5.83 antecedents per rule, and 100% rule accuracy for the Monk3 data set based on the pruned MLP.

Table 7.9. A comparison of results for the Iris data set obtained with different methods.

Methodology	Rule accuracy	Type of decision boundary
Modified RX algorithm based on MLP [145]	97.33%	Hyper-plane
Inputs are transformed into discrete ones artificially based on IMLP [28]	97.33%	Hyper-rectangular
Based on RBF [214]	80%	Hyper-rectangular
Based on RBF [212]	100%	Hyper-rectangular
Our algorithm	100%	Hyper-rectangular

Table 7.10. Rule accuracy and numbers of rules for the Monk3 data set based on DDR.

Rule accuracy	Training accuracy	99.4%
	Validation accuracy	96.6%
	Testing accuracy	98%
The number of premises/rule		3
The number of rules		3

Thyroid Data Set

There are five attributes in the Thyroid data set. There are 215 patterns in the Thyroid data set, 115 patterns for training, 50 for validation, and 50 for testing.

In Chap. 5, $\chi = 0.4$ is selected because it leads to the smallest attribute subset $\{2, 3, 5\}$ with the lowest classification error rates.

Table 7.11 shows the properties of rules for the Thyroid data set. The number of rules is fewer compared to results obtained by GAs (shown in Sect. 7.4), and the rule accuracy is also higher.

Table 7.11. Rule accuracy and numbers of rules for the Thyroid data set based on DDR.

Rule accuracy	Training accuracy	95.65%
	Validation accuracy	95%
	Testing accuracy	95%
The number of premises/rule		3
The number of rules		4

Rules for the Thyroid data set obtained by the present approach are:
Rule 1:
IF attribute 2 is within the interval [11, 25.3]
AND attribute 3 is within the interval [2.9, 10]
AND attribute 5 is within the interval [0, 10.9]
THEN the class label is hyper-thyroid.
Rule 2:
IF attribute 2 is within the interval [15, 23.5]
AND attribute 3 is within the interval [0.77, 2.9]
AND attribute 5 is within the interval [0, 10.2]
THEN the class label is hyper-thyroid.
Rule 3:
IF attribute 2 is within the interval [0.0, 5.12]
AND attribute 3 is within the interval [0.0, 2.12]
AND attribute 5 is within the interval [0, 19.6]
THEN the class label is hypo-thyroid.
Rule 4:
IF attribute 2 is within the interval [0.0, 8.5]
AND attribute 3 is within the interval [0.0, 3.0]
AND attribute 5 is within the interval [14.2, 56.3]
THEN the class label is hypo-thyroid.
Default rule:
the class label is normal.

Breast Cancer Data Set

For the Breast cancer data set, based on the attribute subset $\{2, 3, 7\}$ selected in Chap. 5, we obtain four rules with three antecedents per rule (Table 7.12). The rule accuracy is 97.8% for the test data set. Setiono [286] extracted 2.9 rules and obtained 94.04% accuracy for the Breast cancer data set based on the pruned MLP.

Table 7.12. Rule accuracy and numbers of rules for the Breast cancer data set based on DDR.

Rule accuracy	Training accuracy	96%
	Validation accuracy	97%
	Testing accuracy	97.8%
The number of premises/rule		3
The number of rules		4

The rules for the Breast cancer data set are below:

Rule 1:

IF uniformity of cell shape is within [2, 10]
AND bland chromatin is within [4, 10]
THEN this case is Malignant.

Rule 2:

IF uniformity of cell shape is within [5, 10]
AND bland chromatin is within [2, 10]
THEN this case is Malignant.

Rule 3:

IF uniformity of cell size is within [3, 10]
AND uniformity of cell shape is within [3, 10]
THEN this case is Malignant.

Default rule:

this case is benign.

Mushroom Data Set

There are 22 nominal attributes and 8124 patterns in the Mushroom data set. Among the 8124 patterns, 4500 patterns are for training, 1812 are for validation, and 1812 are for testing.

For the Mushroom data set, based on the attribute subset $\{5, 9, 20\}$ selected in Chap. 5, we obtain 16 rules with the antecedents per rule (Table 7.13). The rule accuracy is 98.86% for the test data set. By the RulEx method [9], four rules, 78 premises, and a rule accuracy of 97.02% were obtained. Setiono [286] extracted three rules with 4.3 premises per rule and obtained 98.12% accuracy for the Mushroom data set based on the pruned MLP. Better rule accuracy is obtained by our method, though the number of rules is larger.

Table 7.13. Rule accuracy and numbers of rules for the Mushroom data set based on DDR.

Rule accuracy	Training accuracy	99.33%
	Validation accuracy	100%
	Testing accuracy	98.86%
The number of premises/rule		3
The number of rules		16

7.6.2 Summary

In this section, we extracted rules from trained RBF neural networks after data dimensionality reduction. Our SCM is used to rank importance of attributes first. According to the ranking results, different attribute subsets are used as inputs to RBF classifiers. The attribute subsets with the lowest classification error rates and the least numbers of attributes are selected. Rules are extracted based on feature subsets selected. Compared to other methods, more concise and accurate rules are extracted for the Iris and Breast cancer data sets. Although, for the Monk3 data set, the rule accuracy is slightly lower, the number of antecedents per rule is smaller than other methods. For the Thyroid data set, compared with the rule method based on GAs in Sect. 7.4, the rule accuracy is higher and fewer rules are needed after data dimensionality reduction. For the Mushroom data set, a high rule accuracy is obtained with fewer premises compared to other methods.

In general, DDR results lead to a less complicated RBF neural network architecture. As a decompositional algorithm, in this rule extraction method, one hidden unit corresponds to one initial rule. Hence, compact rules can be extracted from compact RBF neural networks. Experimental results show that our rule extraction method is simple for implementation and can lead to concise rules and high rule accuracies.

7.7 Rule Extraction Based on Class-dependent Features

7.7.1 The Procedure of Rule Extraction

In this section, rule extraction is carried out for concise rules based on class-dependent features. We demonstrate our approach using computer simulations. The rule extraction algorithm described here is based on the trained RBF neural network classifier with class-dependent features. For each class, a subset of features is selected in order to discriminate the class from other classes. A group of kernel functions is generated for the class based on the selected feature subset. Each hidden neuron of the RBF neural network is responsive to a subset of input patterns (instances).

7.7.2 Experimental Results

The Thyroid data set and the Wine data set from the UCI Repository of Machine Learning Databases [223] are used in this section to demonstrate our algorithm.

Thyroid Data Set

It is shown in the feature masks (Table 6.3) that feature 1 does not play an important role in discriminating classes. Hence, the T3-resin uptake test can be unnecessary in this type of Thyroid diagnosis. For class 3, feature 2 can discriminate class 3 from other classes. Features 2 and 3 are used to

Table 7.14. Rule accuracy for the Thyroid data set based on class-dependent features.

Rule accuracy	Full features	Class-dependent features
Training set	94.57%	95.54%
Validation set	95.35%	94.6%
Testing set	90.7%	95.48%

discriminate class 2 from other classes. Features 2, 3, 4, and 5 are used to discriminate class 1 from other classes.

Two rules are extracted for the Thyroid data set based on class-dependent features. The rule accuracy (in Table 7.14) is 95.54% for the training data set, 94.6% for the validation data set, and 95.48% for the test data set. With full features as inputs, two rules are obtained, and the rule accuracy is 94.57% for the test data set, 95.35% for the training data set, and 90.7% for the validation set. Thus, higher rule accuracy and more concise rules are obtained when using class-dependent features compared to full features.

Rules for the Thyroid data set based on class-dependent features are:

Rule 1:

IF attribute 2 is within the interval (12.9, 25.3)
AND attribute 3 is within the interval (1.5, 10)
THEN the class label is hyper-thyroid.

Rule 2:

IF attribute 2 is within the interval (0, 5.67)
THEN the class label is hypo-thyroid.

Default rule:

the class label is normal.

Wine Data Set

It is shown in the feature masks (Table 6.4) that the feature subset $\{2, 4, 5, 6, 7, 9, 11, 1$ plays an important role in discriminating class 1 from other classes, the feature subset $\{3, 4, 5, 6, 7, 10, 11, 12, 13\}$ is used to discriminate class 2 from other classes, and the feature subset $\{2, 3, 11, 12, 13\}$ is used to discriminate class 3 from other classes.

Seven rules are extracted for the Wine data set based on class-dependent features. The rule accuracy (in Table 7.15) is: 88.7% for the training data set, 83.4% for validation data set, and 86.1% for the test data set. With full features as inputs, seven rules are obtained, and the rule accuracy is 90.6% for the training data set, 77.8% for the validation set, and 86.1% for the test set. Thus, the same rule accuracy and more concise rules with fewer premises are obtained when using class-dependent features compared to using full features.

Table 7.15. Rule accuracy for the Wine data set based on class-dependent features.

Rule accuracy	Full features	Class-dependent features
Training set	90.6%	88.7%
Validation set	77.8%	83.4%
Test set	86.1%	86.1%

7.7.3 Summary

In this section, we have described a rule extraction method from our RBF classifier based on class-dependent features. The discriminatory power of each feature for discriminating classes is considered for each class. Different feature subsets are selected for different classes individually based on their ability in discriminating the class from other classes, which show the relationship between the feature subset and the class concerned. The class-dependent feature selection results obtained above provide a new way for rule extraction. The Thyroid and Wine data sets are used to demonstrate the algorithm. Experimental results show that our algorithm is effective in reducing the number of feature inputs and leads to compact and accurate rules simultaneously.

8

A Hybrid Neural Network For Protein Secondary Structure Prediction

8.1 The PSSP Basics

In this chapter, we will use hybrid neural networks to deal with the protein Secondary Structure Prediction (PSSP) task.

8.1.1 Basic Protein Building Unit — Amino Acid

A *protein sequence* is an array of amino acids, which is called a primary protein structure. Each amino acid is encoded by three out of four DNA bases, i.e., A, C, T, and G. The amino acids are the basic units of protein sequences and are referred to as *residues*. The triplet code implies that there are $4^3 = 64$ possible permutations. However, there are only 20 amino acid types and this results in a redundancy in the genetic code. Thus, almost each of the amino acids (with the exceptions of Methionine and Tryptophan) is encoded by synonymous permutations which are interchangeable in the sense of producing the same amino acid. For convenience of presentation, each amino acid type is represented by an alphabetic letter. For example, the amino acid named Alanine is represented by the letter 'A'. A protein sequence in the alphabetical representation is thus a long sequence of characters, as in the example shown in Fig. 8.1. A protein sequence may be subject to evolutionary changes that may induce mutations, including insertions, deletions, or substitutions, to the original protein sequence and thereafter produce different functions.

8.1.2 Types of the Protein Secondary Structure

Secondary structures are regular structural elements which are formed by hydrogen bonds between relatively small segments of the protein sequence. Often the driving force for the formation of a secondary structure is the saturation of backbone hydrogen donors (NH) and acceptors (CO) with intramolecular

Name: Complex Of Troponin C With A 47 Residue (1-47) Fragment Of Troponin I
PDB ID: 1A2X:B
Sequence:

1) GDEEKRNRAI	TARRQHLKSV	MLQIAATELE	KEEGRREAEK	QNYLAEH
2) GDEEKRNRAI	TARRQHLK _ _	MLQIAATELE	KEEGRREAEK	QNYLAEH
3) GDEEKRNRAI	TARRQHLKSV	MLQIAATELEFFE	KEEGRREAEK	QNYLAEH
4) GDEEKGFRAI	TARRQHLKSV	MLQIAATELE	KEEGRREAEK	QNYLAEH

Note
1) Original Protein Sequence (47 Residues)
2) Deletion: several amino acids deleted from the chain
3) Insertion: amino acids FFE was inserted into the original sequence
4) Substitution: the replacement of amino acids segment by GF

Fig. 8.1. Alphabetical representation of the primary protein sequence and protein mutations

hydrogen bonds. This saturation allows the protein to bury hydrophobic side-chains in its interior (hydrophobic core) without conflicting with the polar backbone. There are three common secondary structures in proteins, namely *α-helix, β-strand* , and *coil.*

α-helix

An α-helix is formed from a connected stretch of amino acids. The α-helix is characterized by hydrogen bonds along the chain, which are almost co-axial. The α-helix is the most abundant helical conformation found in globular proteins. The average length of an α-helix is around 10 residues.

β-strand

A β-strand is the principal component of a β-sheet. The β-sheet is characterized by hydrogen bonds crossing between chains. Each participating β-strand in a β-sheet is not continuous in terms of the primary sequence and does not even have to be close to another β-strand in the sequence. A β-strand has a sequence of 5-10 residues in a very extended conformation.

Coil

Approximately one-third of all residues in globular proteins are contained in coils. The coils in a protein serve to reverse the direction of the polypeptide chain. Coils vary in length.

Fig. 8.2 illustrates the visualized secondary structures of the protein. In the diagram shown, the dark ribbons represent helices. The gray ribbons are β-strands that form the β-sheet. The spring-like strings in between these two secondary structures are the coils that bind them.

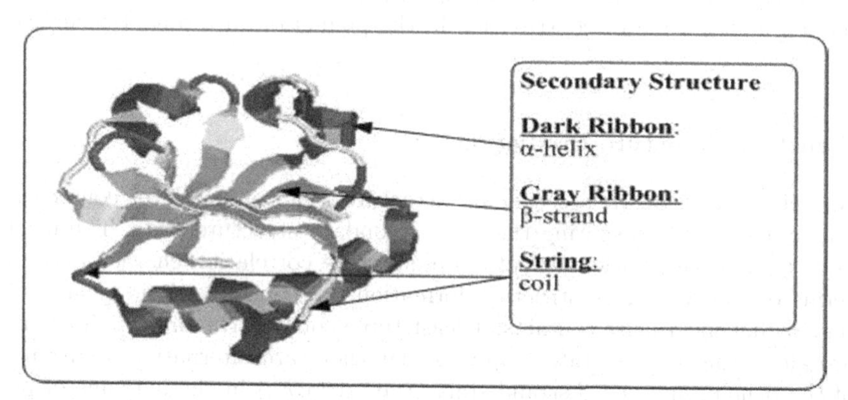

Fig. 8.2. Three types of the protein secondary structure: α-helices are the dark ribbons on the boundary of the diagram, the gray ribbons in the center are the β-strands that form the β-sheet, the coils are the spring-like strings that bind the α-helix and the β-strand.

8.1.3 The Task of the Prediction

The term 'prediction' in the protein secondary structure prediction (PSSP) domain carries a similar meaning as the data mining term 'classification': given a residue of the protein sequence, the predictor should classify it into one out of three secondary structure states based on certain characteristics of the residue. Note that the outcome of the prediction is a state of the secondary structure rather than the secondary structure itself. One residue is only the constitutional element of a secondary structure. A protein secondary structure consists of several residues sharing the same secondary structure state. In other words, the secondary structure state is associated with one amino acid while the secondary structure is for an ensemble of amino acid residues. In the literature, the prediction of the protein secondary structure can be conducted in two stages [164][255][260][262]: the *sequence-structure (Q2T) prediction* and the *structure-structure (T2T) prediction.*

Sequence-Structure Prediction

The Q2T prediction predicts the protein secondary structure from the primary protein sequence of amino acid residues. Given a protein sequence, the Q2T predictor maps each entry of the sequence to the relevant secondary structure state by using the data representation information of the input residue. The data representation attempts to capture information related to the type of the amino acid, the sequence context (that is, what are the neighboring residues of the input), the evolutionary information, etc. The sequence-structure prediction makes a major contribution in the prediction in terms of accuracies.

Structure-Structure Prediction

As defined previously, a secondary structure is an ensemble of consecutive amino acid residues sharing the same secondary structure state. The neighboring sequence positions usually present some correlation characteristics in terms of the secondary structure formation. For example, it is usually observed that an α-helix contains at least three consecutive amino acids which are all in the α-helix state. Suppose that there are alternative occurrences of the α-helix and the β-strand states (i.e., $\alpha\beta\alpha\beta$...) in the outcome of predicted secondary structures; then this prediction must be wrong. The above mentioned example is only a simple example of the correlations existing in the neighboring residues. There are also other correlations that are known or unknown. Therefore, the T2T prediction, which is based on the outputs of the first stage, is needed. This is the second stage of prediction. This stage of prediction attempts to correct unrealistic predictions from the previous stage and thus enhances the overall prediction accuracy. This stage of prediction is the complementary to the sequence-structure prediction.

Fig. 8.3 illustrates the scenario of the secondary structure prediction with two stages of implementations.

It is important to note that the same type of amino acids needs not to be predicted to belong to the same secondary structure state in the secondary structure prediction. For instance, in Fig. 8.3, the 12th and the 20th amino residues counted from the left-hand side are both of type 'F' yet are assigned to two different secondary structure states. It is rather the distinct characteristics embedded within the residue such as the sequence context, the evolutionary information, biochemical properties, speed of translation, etc., that play a more significant role in the formation of the protein secondary structure.

It is acknowledged that the neighboring residues have an impact on the predictive capability of the secondary structure. Therefore, the prediction of the secondary structure at each sequence position should not solely rely on the residue at that position. Rather, a window expanding towards both directions of the residue should be used to include the sequence context. We will discuss the issue in detail later.

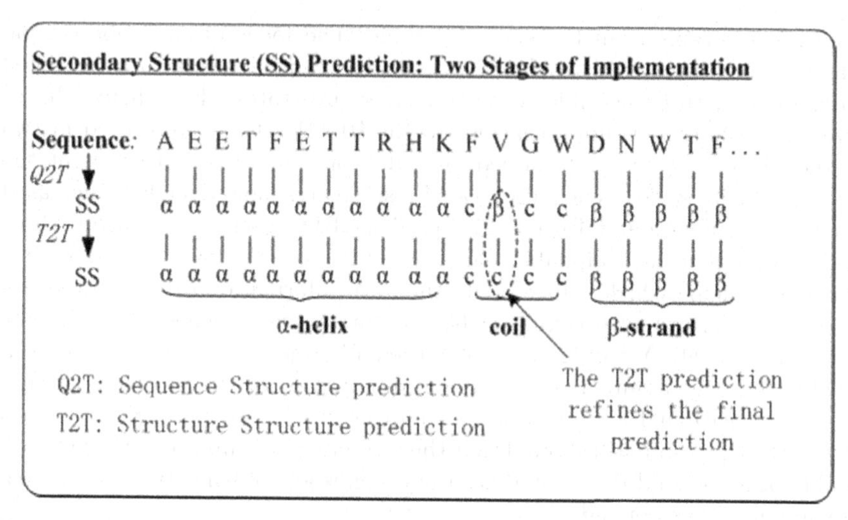

Fig. 8.3. The protein secondary structure prediction: two stages of implementation.

8.2 Literature Review of the PSSP problem

The problem of protein secondary structure prediction was stimulated by the first publication of the protein 3D structure in 1960 [172][238], which attempted to find the correlations between the content of a protein sequence (amino acids) and that of the secondary structure (α-helix, β-strand, and coil). This was when the first generation of secondary structure predictors was born. In this generation, most methods were based on single residue statistics [55][72][194][256][275]. Preferences to particular amino acids from protein sequences were extracted and used in experiments. The accuracies of these methods were found to be over estimated [260].

With the growth of known protein structures obtained from experiments, the second generation exploited segment statistics to predict the secondary structure. A consecutive segment of residues was studied to find how likely the residue of a central segment belonged to a secondary structure state. Major algorithms applied in this generation included statistical information [55][225], sequence patterns [258][302], multi-layer networks [26][140][175][292], multivariate statistics [220], and nearest-neighbor algorithms [271].

The first and the second generations could reach an accuracy level no higher than 70%.

The earliest application of the artificial neural network on protein secondary structure prediction was introduced by Qian and Sejnowski (1988) [248]. In their work, they used a three-layer back-propagation network. The input data was binary encoded by a code scheme called BIN21. Each input data item was a sliding window of 13 residues obtained by extending six sequence positions from the central residue. The focus of each observation was only on the central residue, i.e., only the central residue was assigned to one out of three possible secondary structure states: the α-helix, the β-strand, or the coil. Minor variations to the BIN21 scheme were used in two studies: Kneller *et al.* [175] added one additional input unit to present the hydrophobicity scale of each amino acid residue and showed slightly increased accuracy; Sasagawa and Tajima [272] used the BIN24 scheme to encode three additional amino acid alphabets, B, X, and Z. As all above applied a local encoding scheme which utilized only the local information of the protein sequences, a performance ceiling of 65% accuracy was reached on their own data sets. In 1995, Vivarelli *et al.* [329] used a hybrid system that combined a local genetic algorithm (LGA) and neural networks for the protein structure prediction. Although the LGA has advantages in selecting network topologies efficiently, the result still showed that the ultimate performance of the network could not go beyond the limited accuracy, regardless of what type of network architectures was applied.

More research focusing on analyzing and improving the secondary structure prediction has been carried out [16][68][266][358].

A significant improvement of the 3-state secondary structure prediction accuracy reaching over 70% came from Rost and Sander's work (1993 and 1996) [261][262]. In their method (PHD), a similar multi-layer back-propagation network was used. In contrast to using the BIN21 coding scheme, a new key aspect—the evolutionary information in the form of *frequency statistics from the multiple sequence alignments* was introduced to represent the input data. The inclusion of the protein family information in this form increased the prediction accuracy by around six percent. Moreover, a second cascaded neural network architecture, which performed the additional structure–structure prediction, was used to introduce the correlations between the secondary structure of adjacent amino acid residues in the sliding window. With a data set of 126 protein sequences (RS126), Rost and Sander broke the magic 70% accuracy barrier and achieved an overall network performance as high as 72% in accuracy. Further neural network architecture and machine learning refinements were employed by Riis and Krogh [255]. An adaptive encoding of the input amino acids by the neural network weight sharing technique was used to reduce the number of weights needed. The encoding is adaptive in that the encoding could learn throughout the training process. Specialized networks were designed for each secondary structure class and combined using another neural network. Despite the fact that the architectural design was more complicated, the experiments on the RS126 data set, which again made use of the frequency statistics from the multiple sequence alignment, reached an overall accuracy of 71.3%, which was still similar to that of the PHD method.

More recently, Jones [164] used the *position-specific scoring matrix (PSSM)* [6][291], obtained from the online alignment searching tool PSI-Blast [24], to numerically represent the protein sequence. A PSSM was constructed automatically from a multiple alignment of the highest scoring hits in an initial BLAST search. The PSSM was generated by calculating position-specific scores for each position in the alignment. Highly conserved positions of the protein sequence received high scores and weakly conserved positions received scores near zero. Due to its high accuracy in finding biologically similar protein sequences, the evolutionary information carried by PSSM is more sensitive than those profiles obtained by other multiple sequence alignment approaches. With the neural network architectural similar to that of Rost and Sander's, Jones' PSIPRED method achieved an accuracy as high as 76.5% using a much larger data set than RS126.

In 2001, Hua and Sun [148] also proposed a support vector machine (SVM) approach. This was an early application of the SVM on the PSSP problem. In their work, they first constructed three one-versus-one and three one-versus-all binary classifiers. Three tertiary classifiers were designed based on these binary classifiers through the use of the largest response, the decision tree, and votes for the final decision. By making use of Rost's data encoding scheme, they reached an accuracy of 71.6% and a segment overlap performance of 74.6% on the RS126 data set.

8.3 Architectural Design of the HNNP

As we discussed previously, the secondary structure predictor needs two stages. In this section, we describe the architectural design of a hybrid neural network predictor (HNNP). The predictor consists of two artificial neural networks: a radial basis function (RBF) neural network at its first stage for the sequence–structure (Q2T) prediction and a multi-layer perceptron (MLP) neural network at the second stage for the structure–structure (T2T) prediction.

Figure 8.4 illustrates the overall flow of the HNNP. The 'sequence content' storage contains a number of protein sequences used for the training or the prediction. At each flow through the predictor, only one residue instead of an entire protein sequence is fed as input.

8.3.1 Process Flow at the Training Phase

The flow of the prediction in the training phase is as follows:

1. Given a residue in alphabetic form, it is numerically encoded into a 20-dimensional vector by the 'encoder A'. The 20-dimensional vector contains the frequency statistics obtained from the multiple sequence alignment for the residue.
2. The 20-dimensional vector is then passed to a 'normalizer'. The purpose of the normalization is to restrict values of the vector entries to within the range $(0, 1)$.

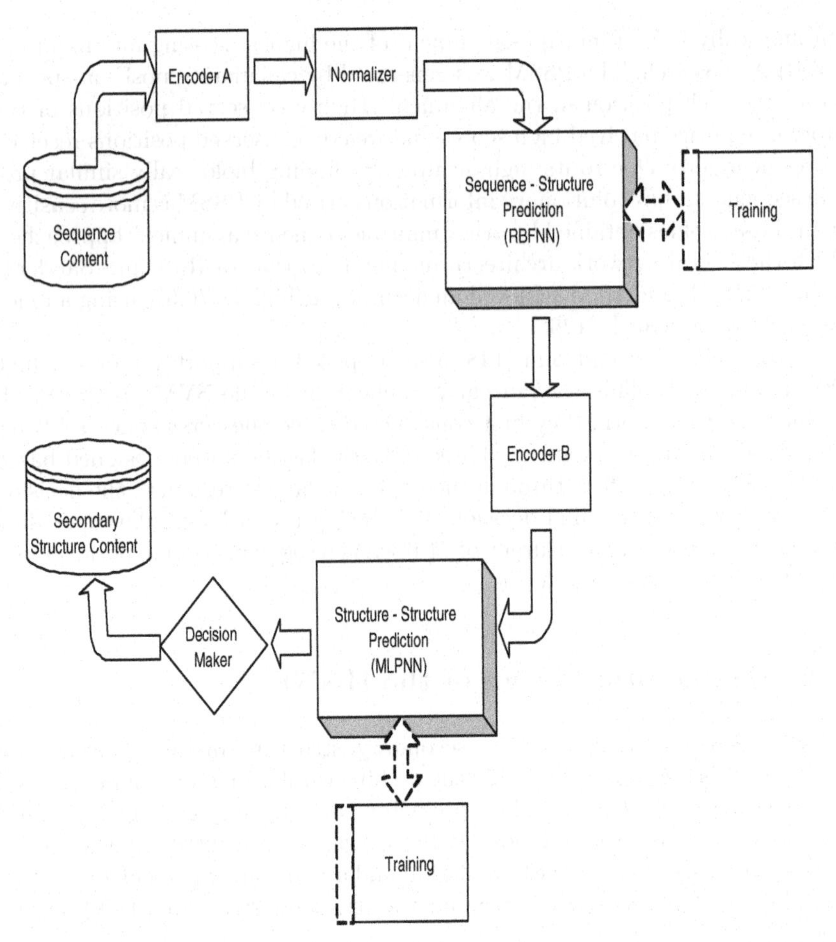

Fig. 8.4. Framework of the hybrid neural network predictor ('MLPNN' stands for 'MLP neural network' and 'RBFNN' stands for 'RBF neural network').

3. The sequence–structure prediction block is a three-layer radial basis function neural network. In this network, the overlapped clustering technique proposed by us [107] was applied to construct the RBF kernels. We have shown in previous chapters that the overlapped clustering technique assisted in enhancing the noise-rejection capability of the classifier and could improve the classification accuracy. In this work, we wish to explore the noise-rejection capability of such an RBF neural network in improving the PSSP accuracy. In the training phase of the sequence–structure prediction, the input was fed into this block to establish the network. This includes the discovering of a set of radial basis function kernels as well as the optimal weight set between the second and the third layers. After the

network is established, the training samples are fed in again one by one. The output is stored as the training data for the next stage.

4. The output from the previous stage is further encoded by the 'encoder B' by real values lying in $(0, 1)$.
5. The structure–structure prediction block is implemented by a three-layer MLP network. The block attempts to correct the predictions made from the sequence–structure prediction and improve the overall prediction accuracy. In the training phase of the structure–structure prediction, the outputs of the first stage are fed into the MLP network one by one. The network learns from these samples and the network parameters are tuned accordingly to minimize the cost function. When the parameters are optimized, the training of the overall predictor is done and testing could be carried on this trained system.

8.3.2 Process Flow at the Prediction Phase

The prediction phase performs the prediction on the testing data set and the results are evaluated for accuracy. Given a protein sequence, the flow of the prediction is as follows:

1. The sequence is properly encoded by the 'encoder A'.
2. The set of 20-dimensional vectors representing the whole protein sequence is normalized.
3. Each protein residue is fed into the sequence–structure prediction block one by one. The RBF neural network uses the established parameters to predict what the relevant secondary structure state is. This prediction generates the primary outcome of the prediction.
4. The output from the previous stage is again encoded by 'encoder B' and fed into the MLP network. The MLP uses the learned parameters to generate the final secondary structure prediction.
5. The 'Decision Maker' uses the winner-take-all (WTA) technique to complete the secondary structure assignment. The secondary structure class with the highest probability is selected to be the predicted secondary structure state. The outcome of secondary structure patterns of a whole sequence is saved in the storage as the 'secondary structure content' of the sequence.

8.3.3 First Stage: the Q2T Prediction

Architecture

The architectural design of the Q2T prediction is shown in Fig. 8.5. The RBF neural network consists of three layers: the input layer, the hidden layer, and the output layer.

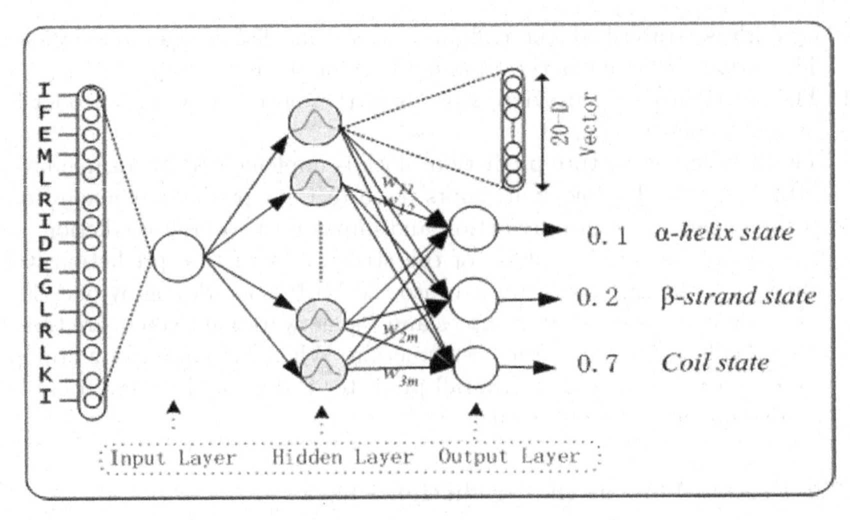

Fig. 8.5. Architecture of the sequence–structure prediction: the first stage.

The *input layer* is where the protein sequence as input is fed into the network. A protein sequence has to be encoded by a certain data representation scheme to allow neural networks to manipulate it. The data representation scheme used will be described later in detail.

The *hidden layer* consists of a set of kernels corresponding to the prototypes in the input space. In this layer, the data is clustered into a finite number of radial regions relevant to the input space. The response of each region can be represented by a kernel function. Typical transfer functions which may be used for these kernels include:

- Gaussian functions;
- Spline functions;
- Multi-quadratic functions;
- Inverse multi-quadratic functions.

In this book, we use the popular Gaussian transfer function of the form:

$$g(x) = \mathrm{e}^{-d(x)^2/2\sigma^2}, \tag{8.1}$$

where x is the input pattern, $d(x)$ is the distance between the input pattern x and the center of the kernel, and σ defines the width of the kernel.

If the input pattern x is one-dimensional, the shape of the Gaussian function is like a bell (Fig. 1.4), which possesses the highest response of 1 at the center and quickly degrades when x moves away from the center. The responses of the kernel for data points whose distances to the center of the kernel exceed twice the kernel width (2σ) are rather weak and may be neglected.

The output layer linearly combines the outputs of hidden neurons. For any output neuron i, the output can be formulated as follows:

$$y_i^{(3)}(x) = \sum_{j=1}^{m} g_{ij}^{(2)}(x) w_{ij}, \tag{8.2}$$

where $g_{ij}^{(2)}(x)$ is the output of the jth kernel for the input pattern x, and w_{ij} is the weight associated with the ith output neuron and the jth hidden kernel. m is the total number of Gaussian kernels (or hidden neurons) in the hidden layer. In our work, each output neuron corresponds to one of the three secondary structure states. The value associated with each neuron indicates the likelihood that the input pattern belongs to that secondary structure state. The decision of the final predicted secondary structure state is made by the winner-take-all (WTA) technique. Assume that the secondary structure states assigned to output units 1, 2, 3 are helix, strand, and coil, respectively. Given an input pattern x, the computed output values of three neurons are 0.1, 0.2, and 0.7, respectively. According to the WTA technique, 0.7 implies the highest likelihood, and the second output neuron wins. The input pattern is thus predicted to be in the secondary structure state associated with the second output neuron – strand.

8.3.4 Sequence Representation

Local Coding Scheme

The sequence of the protein is a series of alphabetical characters. To use it as an input to the neural network, it is necessary to code them into numerical form first before the network could perform manipulations on them. Early work on the PSSP problem used to apply binary coding schemes to fulfill this task. Among them, one typical encoding scheme used is the orthogonal data representation [248], the BIN21. Under this scheme, each amino acid is represented by a 21-dimensional vector. The first 20 entries (bits) of the vector correspond to the occurrences of 20 types of the amino acids at the specific sequence position. If the amino acid to be coded is of type i, then the ith entry of the vector will be coded as 1, while the remaining 19 bits are 0. The additional bit of the vector, the 21st bit, is called a *spacer*. As we mentioned earlier, each input data is a sliding window of length N; there may be cases where the leading or the lagging part of the window may be missing, e.g., the starting residues or the terminating residues of a protein sequence. In this case, the spacer would be able to indicate the completeness of the window.

Position-specific-scoring-matrix (PSSM)

The problem of the BIN21 encoding scheme is that it does not incorporate evolutionary information, which has been shown to provide input domain-specific knowledge in protein secondary structure prediction. To address this problem, a multiple sequence alignment method is applied for the coding of each residue. To give a simple concept of what multiple sequent alignment is, let us consider a pairwise alignment case, without even taking gap insertions into consideration: given two sequences, for instance, 'GDEEKRNRAI' and 'GDSDKNSAAI', they can be aligned as illustrated in Fig. 8.6. Assume that a score of +3 is awarded to one column having residue match (e.g., column 1) and a penalty score of −3 otherwise (e.g., column 3). The aggregated score (6) of all the columns is an *alignment score* raised from this alignment. Note that these two sequences could have many other possible alignments when gaps are allowed to be inserted into sequences and this results in more alignment scores. Among them, the alignment resulting in the highest score is called the *optimal alignment* and this will give a measurement of the similarity between two sequences. While the pairwise sequence alignment is used to determine

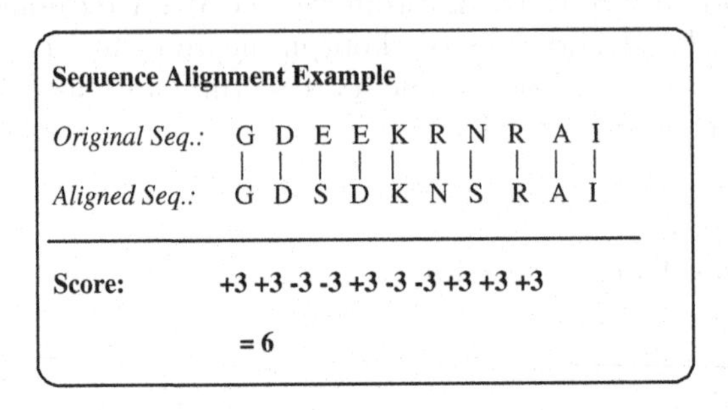

Fig. 8.6. An example illustrating the fundamental sequence alignment theory.

the similarity between two strings, the multiple sequence alignment is used to determine the similarity among multiple strings, i.e., searching for the optimal multiple alignment.

A position-specific scoring matrix (PSSM) is a matrix based on the amino acid frequencies at every position of a multiple sequence alignment. For example, the following three sequences (or segments of protein sequences) have been aligned as shown:
NTEGEWI
NITRGEW
NIGGECC
The frequencies at column 2 for amino acid types T and I are 0.33 and 0.67, respectively. The frequencies for the remaining 18 amino acid types are all 0. However, in order to model every possible sequence, amino acids that do not appear in this column of the multiple alignment will also be considered and assigned a value called a pseudo-count. This results in a 20-dimensional frequency vector with 20 non-zero entries for the column 2 sequence position. Each vector entry corresponds to one of the 20 types of amino acid. Taking the logarithm of each vector entry over the background frequency (e.g., this

may be *a priori* probability of the corresponding amino acid over the entire protein sequence database) gives the final score called *log-odds score*. The log-odds score shows how conservative an amino acid is in its sequence position. Highly conserved positions receive higher scores, weakly conserved positions are scored lower. As each sequence position can be represented by one 20-dimensional vector, the entire protein sequence can thus be represented by a $N \times 20$ *position-specific scoring matrix*, where N is the length of the sequence. Hence, in our example, each of the three protein sequences can be represented by a 7×20 PSSM. More details can be obtained from [6][124][291].

As mentioned earlier, the neighboring residues have an impact on the formation of protein secondary structures. It is important to incorporate this information. For this purpose, we may use a sliding window spanning N residues with the residue of observation located at the center. Thus, each data (residue) is actually represented by an $N \times 20$ matrix. In this sliding window, each window position shares a different degree of importance in determining the secondary structure state. Figure 8.7 illustrates the raw data format for the 10th residue Aspartic acid (D) of the Hydrolase (O-Glycosyl) protein sequence (PDB ID: 119L) using the sliding window size $N = 15$. In this representation, the central residue extends over $(N-1)/2 = 7$ leading/lagging neighboring amino acids and forms a matrix of $N \times 20$ entries. Moreover, one more spacer is added to indicate the completeness of the window. Note that the spacer is only used as the window status indicator and does not contribute to any numerical manipulations regarding the PSSM. In the example given, the spacer 0 implies that both its seven leading and seven lagging residues of the window are available and the window is complete. However, if the focus of observation is on the first residue (position -7) of the window, with sequence alphabet I, the spacer value is -5. This means that for the window centered at that residue, five amino acid residues ($(N-1)/2$ residues – residue 'M' – residue 'N' $= 5|_{N=15}$) are missing and the minus sign indicates that the missing residues are located at the leading part of the window.

Based upon the raw matrix obtained, the data is further normalized to the range $(0, 1)$ by using the standard logistic function:

$$x' = \frac{1}{1 + \mathrm{e}^{-x}}. \tag{8.3}$$

After scaling, the ultimate input data changes to the form as shown in Fig. 8.8.

8.3.5 Distance Measure Method for Data — WINDist

The distance between any two data points may be computed by the Euclidean distance, assuming that each attribute of the vectors shares equal importance. In our clustering problem, however, each data point is represented in the form of a matrix, and each row (i.e., window position) of the matrix may have a

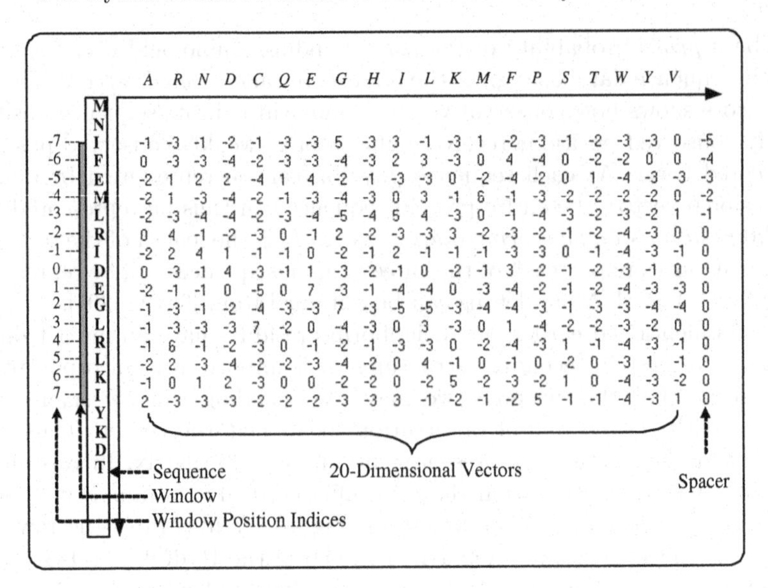

Fig. 8.7. Input data representation: the raw profile (15 × 21) corresponds to the *10*th residue of the Hydrolase protein sequence.

	A	R	N	D	C	Q	...	P	S	T	W	Y	V	
-7 ---	0.26894	0.04743	0.26894	0.11920	0.26894	0.04743	...	0.04743	0.26894	0.11920	0.04743	0.04743	0.50000	-5.00000
-6 ---	0.50000	0.04743	0.04743	0.01799	0.11920	0.04743	...	0.01799	0.50000	0.11920	0.11920	0.50000	0.50000	-4.00000
-5 ---	0.11920	0.73106	0.88080	0.88080	0.01799	0.73106	...	0.11920	0.50000	0.73106	0.01799	0.04743	0.04743	-3.00000
-4 ---	0.11920	0.73106	0.04743	0.01799	0.11920	0.26894	...	0.04743	0.11920	0.11920	0.11920	0.11920	0.50000	-2.00000
-3 ---	0.11920	0.04743	0.01799	0.01799	0.11920	0.04743	...	0.01799	0.04743	0.11920	0.04743	0.11920	0.73106	-1.00000
-2 ---	0.50000	0.98201	0.26894	0.11920	0.04743	0.50000	...	0.11920	0.26894	0.11920	0.01799	0.04743	0.50000	0.00000
-1 ---	0.11920	0.88080	0.98201	0.73106	0.26894	0.26894	...	0.04743	0.50000	0.26894	0.01799	0.04743	0.26894	0.00000
0 ---	0.50000	0.04743	0.26894	0.98201	0.04743	0.26894	...	0.11920	0.26894	0.11920	0.04743	0.26894	0.50000	0.00000
1 ---	0.11920	0.26894	0.26894	0.50000	0.00669	0.50000	...	0.11920	0.26894	0.11920	0.01799	0.04743	0.04743	0.00000
2 ---	0.26894	0.04743	0.26894	0.11920	0.01799	0.04743	...	0.04743	0.26894	0.04743	0.04743	0.01799	0.01799	0.00000
3 ---	0.26894	0.04743	0.04743	0.04743	0.99909	0.11920	...	0.01799	0.11920	0.11920	0.04743	0.11920	0.50000	0.00000
4 ---	0.26894	0.99753	0.26894	0.11920	0.04743	0.50000	...	0.04743	0.73106	0.26894	0.01799	0.04743	0.26894	0.00000
5 ---	0.11920	0.88080	0.04743	0.01799	0.11920	0.11920	...	0.50000	0.11920	0.50000	0.04743	0.73106	0.26894	0.00000
6 ---	0.26894	0.50000	0.73106	0.88080	0.04743	0.50000	...	0.11920	0.73106	0.50000	0.01799	0.04743	0.11920	0.00000
7 ---	0.88080	0.04743	0.04743	0.04743	0.11920	0.11920		0.99331	0.26894	0.26894	0.01799	0.04743	0.73106	0.00000

Fig. 8.8. Input data representation: the normalized profile.

different degree of importance. Therefore, a data specific distance measure should be used. Here, we propose a unique distance measure WINDist, which is based on the Euclidean distance and takes the data characteristics into consideration. To compute the distance between two data points, the basic idea of WINDist is as follows: given two data points (matrices) A and B, we first match the two data points row by row, e.g., the first row of matrix A matches that of matrix B while the last row of matrix A matches that of matrix B. We then compute the distance between the two row vectors using the Euclidean distance and finally use the weighted sum of those row distances as the final distance measure.

What follows is a more detailed description of WINDist. Denote the ith data point of a given sequence in the following form:

$$D_i = [r^i_{-(N-1)/2}, r^i_{-(N-1)/2+1}, ..., r^i_0, ..., r^i_k, ..., r^i_{(N-1)/2-1}, r^i_{(N-1)/2}], \quad (8.4)$$

where r^i_k is a 20-dimensional vector corresponding to the kth window position, and k is either a positive or a negative integer in the range $[-(N-1)/2, (N-1)/2]$. To compute the distance of D_i to another data point D_j, we first calculate the distance between any two matching rows using the Euclidean distance as follows:

$$d^{i,j}_k = \parallel r^i_k - r^j_k \parallel, \quad (8.5)$$

where $\parallel \cdot \parallel$ denotes the Euclidean norm. As the central position (represented by r^i_0) makes the most important contribution to the prediction while the importance of the neighboring ones degrades with position, we assign a weighting factor to each position. If $k = 0$, it means that this corresponds to the central position and the weight associated with it is the most important, i.e., a weighting factor of 1. For the other positions, we assume that *the significance of these residues degrades with their displacements to the central position.*

For ease of explanation, let us first define a term $p(ab)$ addressing the relative displacement of two window positions with corresponding residue appearances a and b, respectively:

$p(ab)$ = displacement of b to central residue − displacement of a to central residue.

From the definition, it is obvious that the p value is a relative difference of 'closeness' to the position at the center. In Fig. 8.9, the p value $p(E, L) = 4$ for residues E and L means that E is four positions closer to the central residue than L. Now we shall find a way to determine the weighting factor associated with each row of the matrices. Assume that the degrading of the significance of each row vector is in a unique ratio b, which we call the base. The base value should be within the range [0, 1]. If the base value is 1, it is equivalent to the case where no weighting is used and all window positions are equally important. We propose the two alternative weighting schemes:

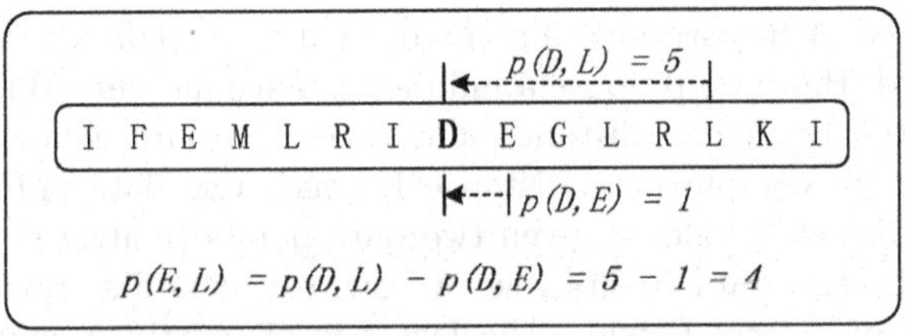

p (D, L) = 5

I F E M L R I **D** E G L R L K I

p (D, E) = 1

p (E, L) = p (D, L) − p (D, E) = 5 − 1 = 4

Fig. 8.9. Relative displacement of the residues.

1. Scheme A: unsupervised weighting scheme. The weighting factor degrades in a geometric proportion with respect to the displacement. In other words, the weighting factor associated with residue r could be formulated as
$$w_r = b^{|p(r,c)|}, \tag{8.6}$$
where $p(r,c)$ is the displacement of the residue r with respect to the central residue c.
2. Scheme B: supervised weighting scheme. For two residues r^1, r^2 with $p(r^1, r^2) = 1$, the weighting factor degrades if and only if the corresponding secondary structures are different. The reasoning for this weighting scheme is as follows: the secondary structure of a protein is usually formed by some consecutive amino acid residues. The consecutive residues sharing the same secondary structure states may contribute *equally* to the formation of the secondary structure and hence are of equal importance. Figure 8.10 illustrates how the weighting factors are assigned for the given sequence and secondary structure pair, 'IFEMLRIDEGLRLKI' and 'HHHHHEEEEHHHHHC', respectively.

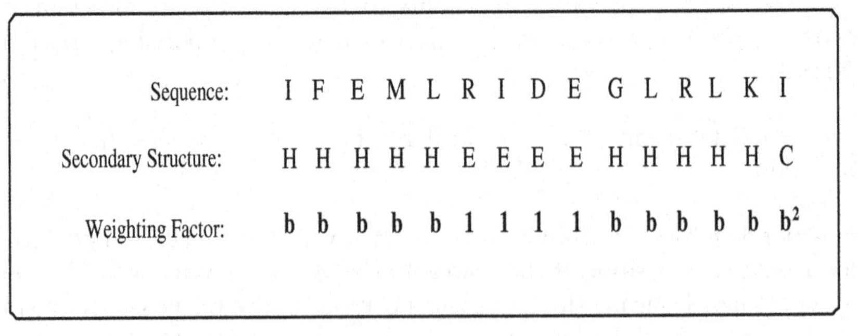

Fig. 8.10. Supervised weighting factor assignment: scheme B

With the Euclidean distances computed for each row vector, the aggregation of all these distances from the N residue positions gives the final estimation of the distance measure:

$$d(r^i, r^j) = \sum_{k=-(N-1)/2}^{(N-1)/2} (w_k \parallel r_k^i - r_k^j \parallel) / \sum_{k=-(N-1)/2}^{(N-1)/2} w_k, \tag{8.7}$$

where w_k is the weighting factor obtained by scheme A or scheme B.

Up to now, our distance measure deals with the situation that both the data points have a full sliding window. However, there are situations where the sliding window might not be full, e.g., the starting or the terminating $(N-1)/2$ residues of the sequence. In this case, we only consider the overlapped part of two windows. Equation (8.7) is then generalized to the following form:

$$d(r^i, r^j) = \sum_{k=-l}^{h} (w_k \parallel r_k^i - r_k^j \parallel) / \sum_{k=-l}^{h} w_k, \tag{8.8}$$

where l is the number of overlapped positions in the leading half-window and h is the number of overlapped positions in the lagging half-window. For a complete window, both the numbers of overlapped leading and lagging vectors are equal to $(N-1)/2$ and hence Eq. (8.7) is a special case of Eq. (8.8).

8.3.6 Second Stage: the T2T Prediction

Architecture

The second stage of our predictor is a three-layer MLP neural network. Similar to the RBF neural network, the network consists of an input layer, a hidden layer, and an output layer. Although the back-propagation technique allows any number of layers of the network to be trained, a three-layer structure is sufficient for approximating any function with finitely many discontinuities to an arbitrary precision, provided that the activation functions of the hidden units are non-linear (the universal approximation theorem) [69][113][131][144]. The architectural layout of the network is shown in Fig. 8.11.

The *input layer* is where the windowed data are fed in. The number of neurons required depends on the number of numerical entries in the windowed data and is $N \times L$ in our problem, where N is the window size and L is the vector dimension.

The input to the *second layer* is the weighted sum of the input data:

$$s_k^{(l)} = \sum_{i=1}^{m_{l-1}} (w_{ki}^{(l)} y_i^{(l-1)} + \theta_{l-1}), \tag{8.9}$$

where $w_{ki}^{(l)}$ is the weight connecting the kth neuron in layer $l = 2$ and the ith neuron in layer $l-1 = 1$ and θ_{l-1} is the bias feedfowarded from layer $l-1 = 1$. $y_i^{(l-1)}$ is the output of the ith neuron in layer $l-1$, i.e., the input layer. m_{l-1} is the number of the neurons in layer $l-1$.

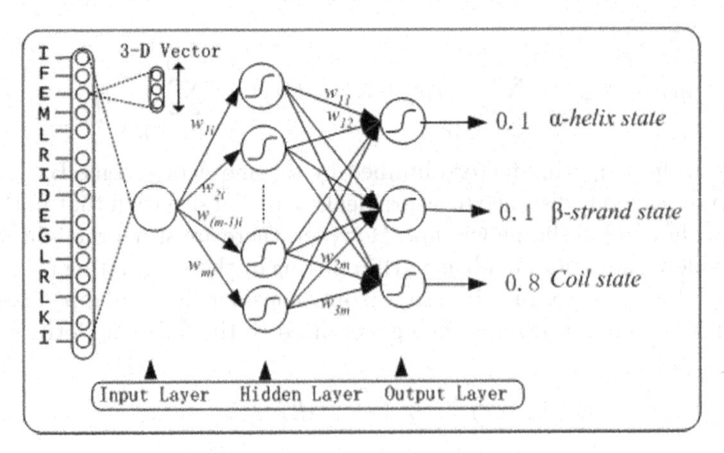

Fig. 8.11. Architecture of the structure–structure prediction: the second stage.

With the input computed, the output or the activation of the neuron is obtained by a sigmoid function:

$$y_k^{(l)} = \frac{1}{1 + e^{-s_k^{(l)}}}, \tag{8.10}$$

where $y_k^{(l)}$ is the activation of $s_k^{(l)}$ in the kth neuron of the lth layer. The behavior of a sigmoid function with scalar input x is shown in Fig. 8.12.

Transformed by the sigmoid function, the inputs with large positive values are activated closely to 1 while large negative ones are activated to 0. For the inputs close to 0, the responses are linearly related to the input values. The outputs of the second-layer neurons are thus smoothly restricted to the range $(0, 1)$.

The *output layer* functions exactly the same as the hidden layer. The input to this layer is the weighted sum of outputs of the neurons from the preceding layer. A sigmoid function is then used to smoothly restrict the output to a desirable range.

By comparing the RBF neural network and the MLP neural network, it can be seen that the networks are quite similar. The data is propagated layer by layer, from the input layer to the output layer. Each layer performs some transformation to the data before it is forwarded. The output layer constitutes three neurons, each corresponding to one secondary structure class.

Despite the similarities mentioned above, the MLP network still possesses some variations in several aspects:

- The input neuron of the MLP neural network as in Fig. 8.11 is an assembly of $N \times L$ subneurons, with each subneuron connecting to all the neurons of the hidden layer. The capital letter L here represents the dimensionality of the vector of each residue.

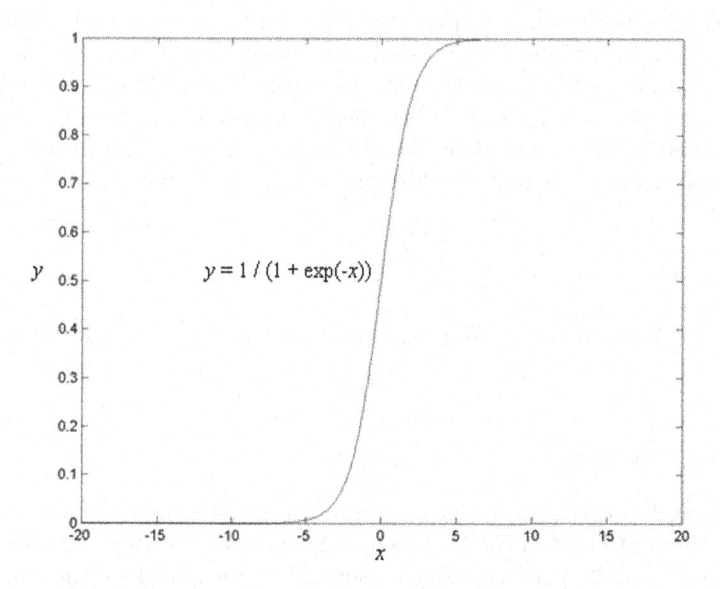

Fig. 8.12. Response of the sigmoid function.

- Weights between the input layer and the hidden layer neurons are adjusted during training. For the RBF neural network, all weights between these two layers are fixed to be 1.
- The activation function used in the hidden layer of an MLP neural network is different from that of the RBF neural network.
- In the RBF neural network, the weighted sum of the activations from the hidden layer is directly used as the output. In the MLP neural network, the additional smooth limiting is performed by the output neurons using the sigmoid function. In other word, the output neurons in the RBF neural network are linear, whereas the output neurons in the MLP neural network are non-linear.

Due to the architectural difference, the encoding of the data is slightly different from that of RBF neural network.

8.3.7 Sequence Representation

Input preprocessing

The output of the first stage is a three-dimensional vector, with each entry representing how likely the residue in observation belongs to the corresponding secondary structure class. In other words, the outputs of the RBF neural network could be interpreted as the probabilities that the residue belongs to the three classes. Thus, it is expected that the values are to be restricted within $(0, 1)$ and the sum of them shall equal to 1. Besides, it is also desirable that the data fed into the MLP neural network is in that range. Therefore, a preprocessing of the data is performed before the outputs of the first stage are directly used as the inputs of the MLP network. To satisfy these constraints, the normalized exponential transformation technique softmax [31] is used:

$$Q_j(x) = \frac{\mathrm{e}^{V_j(x)}}{\sum_{k=1}^{3} \mathrm{e}^{V_k(x)}}, \tag{8.11}$$

where $V_j(x)$ is the jth entry of the output vector and $Q_j(x)$ is the transformed value.

Encoding Scheme

Considering the correlations existing between neighboring residues, a sliding window similar to that in the RBF neural network also needs to be used in the MLP neural network. As mentioned, there may be cases where the window is not complete. On the other hand, the input layer of the proposed MLP neural network architecture consists of a fixed number: $N \times L$ neurons, which means that the network could only accept the data with a full sliding window. To resolve the conflict, some modifications have to be done to the data before they could actually be used. Similar to the case of the RBF neural network, a *spacer* is used as an additional entry to the three-dimensional vector. The spacer is an indicator of the window status. For each window, if the vector corresponding to the window position contains realistic values (e.g., has a sequence residue at that position), the spacer is set to 0. Otherwise, the spacer is set to 1. In the case where the window position is not realistic, artificial values would have to be filled. Here, we adopted the scheme used by Rost and Sander [262] in their work: all the entries where data are not available in incomplete windows are set to 0 except for the spacer. Figure 8.13 shows the encoded form of the data for (a) a full-window case and (b) a partial-window case. A window size of $N = 15$ is used in the illustration.

Similar to that in the RBF network, the data should also be expanded on its neighborhood before it can be used to set up the network or carry out the prediction.

The training of the hybrid neural network predictor is composed of two parts, i.e., the training of an RBF neural network and the training of an MLP neural network, which will follow the work stated in previous chapters.

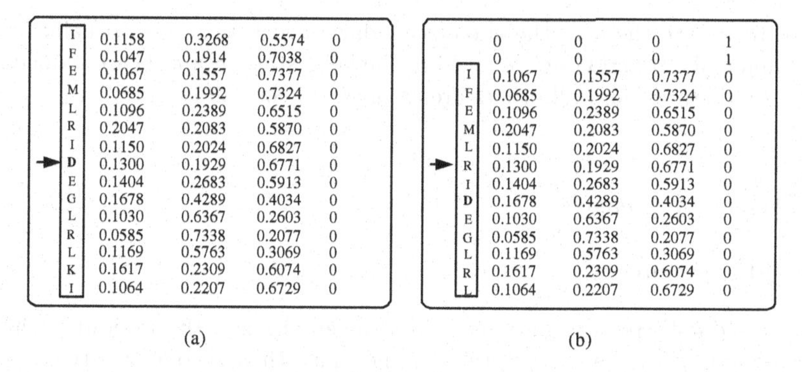

(a)

I	0.1158	0.3268	0.5574	0
F	0.1047	0.1914	0.7038	0
E	0.1067	0.1557	0.7377	0
M	0.0685	0.1992	0.7324	0
L	0.1096	0.2389	0.6515	0
R	0.2047	0.2083	0.5870	0
I	0.1150	0.2024	0.6827	0
D	0.1300	0.1929	0.6771	0
E	0.1404	0.2683	0.5913	0
G	0.1678	0.4289	0.4034	0
L	0.1030	0.6367	0.2603	0
R	0.0585	0.7338	0.2077	0
L	0.1169	0.5763	0.3069	0
K	0.1617	0.2309	0.6074	0
I	0.1064	0.2207	0.6729	0

(b)

	0	0	0	1
	0	0	0	1
I	0.1067	0.1557	0.7377	0
F	0.0685	0.1992	0.7324	0
E	0.1096	0.2389	0.6515	0
M	0.2047	0.2083	0.5870	0
L	0.1150	0.2024	0.6827	0
R	0.1300	0.1929	0.6771	0
I	0.1404	0.2683	0.5913	0
D	0.1678	0.4289	0.4034	0
E	0.1030	0.6367	0.2603	0
G	0.0585	0.7338	0.2077	0
L	0.1169	0.5763	0.3069	0
R	0.1617	0.2309	0.6074	0
L	0.1064	0.2207	0.6729	0

Fig. 8.13. Two examples of encoded form of the data: (a) a full window and (b) a partial window.

8.4 Experimental Results

8.4.1 Experimental Data set

In the PSSP, if an unknown protein sequence shares a significant degree of sequence similarity with the known protein sequence or is a homology, the structure of the unknown protein can be reliably predicted by the simple comparison and the reference between the two sequences [6][53][54][178][277][278]. On the other hand, it is not always true in practice that a new sequence has its homology whose structure has been identified. Hence, to evaluate prediction performance fairly, the bias arisen from homology has to be removed. Moreover, according to Cuff and Barton [68], sequences with low sequence similarity yet significant structural similarity can remain by pairwise sequence alignment methods. Structural similarity also needs to be investigated so as to further remove the bias towards producing favorable prediction accuracy.

The data set used here was originally developed and used by Jones [164] for training and testing his PSIPRED method. The data can be downloaded from the website http://bioinf.cs.ucl.ac.uk/psipred/. The PSIPRED data set contains 2235 protein sequences in total for the training set and 187 sequences for the testing set. The data set was derived by considering constraints for the fair prediction evaluation, as discussed earlier. All the 187 test sequences have been scanned against the training sequences to ensure that they did not share the similar fold. Jones believed that such a practice is at least equivalent to the 7-fold cross-validation test and the testing result should not be over estimated [164]. With such specially designed training and test data sets, no further 7-fold cross validation [16][262] is needed to avoid the over-estimation.

As the secondary structures given in the data set consist of eight states: H, I, G, E, B, S, T, -, and the scheme outlined by Rost and Sander [262] was adopted to reduce them to the three secondary structure states (α-helix, β-strand, and coil). H and G are reduced to the helix state. E and B are reduced to the strand states, and all others are assigned to the coil state.

8.4.2 Accuracy Measure

Per-residue Accuracy

The per-residue-based accuracy measurement emphasizes the fraction of individual residues that are correctly classified into their secondary structure states. The measure could either return an aggregate accuracy over all sequences or an average of the accuracies of all individual protein sequences. What follows is a mathematical description of the way the accuracy is derived. Suppose that we have a protein sequence k and define

$$a_{ij} = \text{number of residues predicted to be in secondary structure state } j \text{ and observed to be in state } i, \tag{8.12}$$

where $i, j \in \{1 = helix, 2 = strand, 3 = coil\}$. The sum of a_{ij} for i from 1 to 3 gives rise to the number of residues predicted to be in class j:

$$P_j^k = \sum_{i=1}^{3} a_{ij}. \tag{8.13}$$

The sum of a_{ji} for i from 1 to 3 gives rise to the number of residues observed to be in class j:

$$O_j^k = \sum_{i=1}^{3} a_{ji}. \tag{8.14}$$

The sum of all residues in the data set is

$$N^k = \sum_{i=1}^{3}\sum_{j=1}^{3} a_{ij}. \tag{8.15}$$

For class $r \in \{\alpha = helix, \beta = strand, c = coil\}$, the accuracy achieved is thus computed as:

$$Q_r^k = \frac{a_{jj}}{O_j}, \tag{8.16}$$

where r, j are the indices representing the same class. The overall accuracy of the prediction for protein sequence k, denoted as Q_3^k, is the aggregated sum of the correct predictions over the total number of residues in the sequence:

$$Q_3^k = \frac{\sum_{j=1}^{3} a_{jj} N^k}{N}. \tag{8.17}$$

Hence, if the data set contains p protein sequences with N amino acid residues in total, the prediction accuracy Q_3 for all the p sequences is:

$$Q_3 = \frac{\sum_{k=1}^{p} Q_3^k N^k}{N}. \tag{8.18}$$

The average of the accuracies for all individual sequences can be computed as

$$Q_{3s} = \frac{\sum_{k=1}^{p} Q_3^k}{p}. \tag{8.19}$$

Matthews' Correlation

A more complicated accuracy measure is to use Matthews' correlation [210] which evaluates the success of predicting residues for each secondary structure state. The formula for computing Matthews' correlation is given by:

$$C_i = \frac{p_i n_i - u_i o_i}{\sqrt{(p_i + u_i)(p_i + o_i)(n_i + u_i)(n_i + o_i)}}, \tag{8.20}$$

for $i \in \{\alpha, \beta, c\}$. Here p_i is the number of correctly predicted residues in the secondary structure state i; n_i is the number of those correctly not assigned to the state i; and u_i is the number of underestimated and o_i is the number of overestimated predictions. Mathematically, these terms can be expressed as

$$\begin{aligned} p_i &= a_{ii}, \\ n_i &= \sum_{j \neq i}^{3} \sum_{k \neq i}^{3} a_{jk}, \\ o_i &= \sum_{j \neq i}^{3} a_{ji}, \\ u_i &= \sum_{j \neq i}^{3} a_{ij}, \end{aligned} \tag{8.21}$$

for $i \in \{\alpha, \beta, c\}$.

Segment-Overlap Measure

The segment overlap measure (SOV) was first proposed by Rost *et al.* in 1994 [263] and later modified and improved by Zemla *et al.* in 1999 [355]. This scheme gives a measure of how well the predicted secondary structure segments are matched with the observed (obtained from experiments) ones. The accuracy is computed by counting the predicted and the observed segments and measuring their overlaps. Let (s_1, s_2) denote a pair of overlapping segments. Define set $S(i)$ as

$$S(i) = (s_i, s_2) : s_1 \cap s_2 \neq \oslash; \tag{8.22}$$

represent the set of all overlapping pairs of segments (s_1, s_2). s_1 and s_2 are both in the secondary structure state i.

Another set, which defines the set of all segments (s_1, s_2) sharing no overlapping in their segment extents, is defined as

$$S'(i) = s_1 : \forall s_2, s_1 \cap s_2 = \oslash. \tag{8.23}$$

The segment overlap measure for state i is thus defined as

$$SOV(i) = 100 \times \frac{1}{N(i)} \sum_{S(i)} \left[\frac{min_{ov}(s_1, s_2) + \delta(s_1, s_2)}{max_{ov}(s_1, s_2)} \times len(s_1) \right], \tag{8.24}$$

with the normalization value

$$N(i) = \sum_{S(i)} len(s_1) + \sum_{S'(i)} len(s_1). \tag{8.25}$$

The sum in Eq. (8.24) and the first sum in Eq. (8.25) are computed from all the segment pairs in state i which overlap by at least one residue, the second sum in Eq. (8.25) is computed from the remaining segments in state i found in the observed (or target) assignment, $len(s_1)$ is the number of residues in segment s_1, $min_{ov}(s_1, s_2)$ is the length of the actual overlap of s_1 and s_2, i.e., for which both segments have residues in state i, $max_{ov}(s_1, s_2)$ is the total extent for which either of the segments s_1 and s_2 has a residue in state i, and $\delta(s_1, s_2)$ is defined as:

$$\delta(s_1, s_2) = min \begin{cases} max_{ov}(s_1, s_2) - min_{ov}(s_1, s_2), \\ min_{ov}(s_1, s_2), \\ int(len(s_1)/2), \\ int(len(s_2)/2). \end{cases} \tag{8.26}$$

Generalization based on Eq. (8.24) results in a 3-state SOV measurement:

$$SOV(i) = 100 \times \left[\frac{1}{N} \sum_{i \in [\alpha, \beta, c]} \sum_{S(i)} \frac{min_{ov}(s_1, s_2) + \delta(s_1, s_2)}{max_{ov}(s_1, s_2)} \times len(s_1) \right], \tag{8.27}$$

where the normalization value N is the sum of $N(i)$ over all three secondary structure states.

Q2T Prediction — the Raw Prediction

As we discussed earlier, the prediction of the secondary structure is implemented in two stages: the sequence–structure prediction and the structure–structure prediction. For ease in implementation, training of the two stages is performed independently. Upon the accomplishment of the training, the test data may then be passed through the stages of the system. The output from the second stage determines the final secondary structure state assignment.

The Q2T prediction is implemented by an RBF neural network. To obtain the optimal performance of the prediction, several parameters related to the network topology, the kernel formation, etc., have to be selected through experiments. More precisely, these parameters include

1. The base ratio b. This term determines the degrading of the weighting factors in the proposed distance measure WINDist as described previously.
2. The alternative WINDist scheme A or B. Both schemes are proposed on the basis of some assumptions. Experimental results have to be used to find the better scheme.
3. The window size N. A small window size may lose useful information associated with farther-neighboring residues. On the other hand, bringing residues too distant away into observation may introduce noise, as the information from those residues is usually irrelevant. We need to find a suitable window size which exploits a sufficient amount of relevant information without introducing noise.
4. Purity level ρ of the clusters (kernels). A higher ρ value results in a more accurate response from the kernel for training data, but may result in lower generalization performance for test data.

Ideally, these parameters should be tuned together such that a combination which leads to the optimal performance of the Q2T prediction could be found. However, doing so would demand a huge amount of computational burden and time, especially when the data used in our work is large in scale and high in dimensionality. According to our simulations, each trial of training takes 1–2 days to finish by the computational resource available to us (a 32-CPU Linux cluster, 3GHz, with 2GB RAM for each CPU). For this reason, we adopted a relative suboptimal yet time-saving tuning scheme: change some sets of parameters, fix the rest.

Now, we shall determine which parameters are to be tuned first. Basically, the training of the RBF network starts by applying the overlapped clustering algorithm. The parameters affecting this early step are the base ratio b and the WINDist scheme. Both variables would affect the distance measure between data points and hence influence the clustering. Thus, these two variables are to be tuned first.

8.4.3 Experiments with the Base and Alternative Distance Measure Schemes

Strategy

To determine a suitable base ratio b and the best WINDist scheme, we run simulations on two WINDist schemes (A, B) independently. For each WINDist scheme, the base value b is varied from 0.3 to 0.7 with an increment of 0.1 each time. The purpose of doing so is first to find the best base value for each scheme and then select the scheme which outperforms the other.

Results

Table 8.1 shows the experimental results.

Table 8.1. Performance comparison of two WINDist schemes: the base value b is varied within the range [0.3, 0.7] with an increment of 0.1 each time

Base	**Scheme A**				**Scheme B**			
(b)	Q_3(%)	Q_α(%)	Q_β(%)	Q_c(%)	Q_3(%)	Q_α(%)	Q_β(%)	Q_c(%)
0.3	*74.0*	74.9	56.6	82.7	*74.6*	75.6	57.7	82.8
0.4	*74.5*	75.5	56.5	83.3	*74.9*	75.8	58.0	83.4
0.5	*74.7*	75.6	56.9	83.5	*75.2*	76.2	57.9	83.7
0.6	*74.4*	75.3	56.4	83.3	*75.1*	76.0	57.9	83.6
0.7	*74.2*	75.4	56.1	82.8	*74.6*	76.1	56.8	82.5

We can see by observing the table that the general performance of scheme B is superior to that of scheme A. The highest Q_3 accuracy that scheme B could reach is 75.2% when the base value is set to 0.5. It happens that the highest Q_3 of scheme A is also reached at the same base yet the accuracy is only 74.7%. This accuracy is lower than that of scheme B. Comparison of Q_3 values at other base values also shows that scheme B is superior to scheme A. Thus, it is believed that the supervised WINDist Scheme is more appropriate to our system and that all further simulations would use this scheme.

Besides, by observing the Q_3 column of scheme B, we can see that base values either larger or smaller than 0.5 degraded the prediction accuracies. Expectedly, this base value has more relevantly reflected the degrading of the significance of the neighboring residues and is more relevant to our proposed WINDist scheme.

8.4.4 Experiments with the Window Size and the Cluster Purity

Strategy

By deciding the optimal value for the base and the appropriate scheme for the distance measure in advance, we have effectively reduced the computational demand such that it is affordable to us. Now we are left with two major parameters:

- The window size which determines the span of the sliding window.
- The cluster purity which decides the minimum purity that a cluster (kernel) has to satisfy.

To get the best performance of the first stage, we have tried different combinations of these two parameters. To carry out the simulations in an orderly way, the following strategy is adopted:

1. Define some possible starting values of both parameters. For the window size, we started with a value of 15, which is within the range of the popular window sizes [7, 17] [348]. The recent predictors like PHD, PSIPRED, etc., have already reached an accuracy level higher than 70%; considering that the contribution of a kernel should also be reliable enough to the predictor, we selected a purity level of $\rho = 0.6$ as the starting point.
2. Define a step size for each parameter. For a window size N, a minimum step size of 2 is used; this is equivalent to expanding one position further towards the both sides of the neighborhood. For the purity level, the step size is set to $\Delta\rho = 0.05$ (except the change of the purity level from 0.6 to 0.7, as the accuracies obtained from the initial purity level are relatively low).
3. Run simulations on the initial parameter set.
4. Keep the window size intact and increase the purity level step by step until it reaches 1.0. Conduct experiments at each step.
5. Increase the window size N by 2. Repeat step 4.
6. Repeat step 5 until a significant performance degrading is observed.
7. Reset the window size to the starting point. Decrease the window size step by step and repeat experiments with respect to ρ at each step. Stop further trials if a significant decay in performance is observed.
8. Record the accuracies returned from all trials of simulation.

By conducting experiments in the above strategy, we could see how the generalization performance of the prediction of the RBF neural network is impacted by the purity value ρ. Meanwhile, we would also be able to tell which window size is optimal to the prediction of the RBF neural network.

Results

Table 8.2 shows the experimental results for window size $N = 11$ with the purity level ρ changing from 0.6 to 1.0. The Q_3 column of the table shows the overall accuracies obtained with different purity levels. From the column, we see that the change of accuracy against the purity includes two parts. In the first part, when ρ keeps increasing from 0.6 through 0.85, the accuracy is either increasing or maintained at the top accuracy value. This is where the higher cluster purity plays an 'active' role: the higher purity level makes each cluster more accurate in response and hence improves the prediction accuracy. Also, higher cluster purity results in better noise rejection. When the purity goes

Table 8.2. Prediction accuracies of the sequence–structure prediction with window size $N = 11$ and various purity levels ρ.

Window size (N)	**Purity level (ρ)**	**Accuracies**			
		Q_3(%)	Q_α(%)	Q_β(%)	Q_c(%)
11	**0.6**	*72.9*	73.9	55.1	81.6
	0.7	*74.7*	75.5	57.1	83.4
	0.75	*74.8*	75.6	57.2	83.5
	0.8	*74.8*	75.6	57.3	83.5
	0.85	*74.8*	75.6	57.3	83.4
	0.9	*74.7*	75.3	57.3	83.6
	0.95	*74.6*	75.1	57.2	83.5
	1.0	*74.3*	74.9	56.7	83.1

beyond 0.85 and increases further to 1.0, this is the second part of the change. In this part, any further increase in ρ reduces the generalization capability of the RBF network and this results in the drop of the prediction accuracy.

Tables 8.3, 8.4 and 8.5 show the experimental results for window size $N \in \{13, 15, 17\}$, $\{19, 21, 23\}$, and $\{25, 27\}$, respectively. In these results, a similar scenario of the accuracy change against the purity may be observed. On the other hand, the ρ values for peak accuracies are not exactly the same for different window sizes. This variation is reasonable: the window size determines the span of the window; for various window sizes, the information carried by the data may be different; consequently, the distance computed for any two data points would also change, and this, in turn, influences the value of the optimal purity level ρ for clustering. Despite the difference in optimal purity value ρ, what important is that there is always a ρ balancing the accurate responses of the kernels and the good generalization performance of the RBF neural network.

To see the performance of the RBF prediction with different window sizes, we summarize the best accuracies obtained for each window size in Table 8.6. From this table, the relationship of the window size N and the prediction accuracy of the RBF neural network is quite obvious. The best Q_3 accuracy is 75.6% when the window size is $N = 21$. A window size either larger or less than 21 leads to the degraded prediction performance. In other words, the window size 21 has most relevantly captured the useful information hidden in the neighborhood and serves best for the learning of our RBF neural network.

8.4.5 T2T Prediction — the Final Prediction

Feasibility of the Back-Propagation

With the optimal prediction performance reached at the first stage, we should continue optimizing the second stage such that the overall prediction capabil-

Table 8.3. Prediction accuracies of the sequence-structure prediction with various window sizes N and purity levels ρ: $N \in \{13, 15, 17\}$.

Window size (N)	Purity level (ρ)	Accuracies			
		Q_3(%)	Q_α(%)	Q_β(%)	Q_c(%)
13	**0.6**	*73.4*	74.3	56.0	81.9
	0.7	*74.9*	76.2	57.6	83.2
	0.75	*75.1*	75.9	57.7	83.7
	0.8	*75.0*	75.6	57.4	83.8
	0.85	*75.1*	76.0	57.6	83.8
	0.9	*75.0*	75.7	57.5	83.8
	0.95	*74.9*	75.9	57.2	83.7
	1.0	*74.8*	75.8	57.2	83.5
15	**0.6**	*73.4*	74.4	56.2	81.8
	0.7	*75.2*	76.2	57.7	83.6
	0.75	*75.2*	76.2	57.9	83.7
	0.8	*75.2*	76.2	57.8	83.8
	0.85	*75.2*	76.1	57.7	83.8
	0.9	*75.3*	76.2	58.2	83.7
	0.95	*75.2*	75.8	58.2	84.0
	1.0	*75.0*	75.5	57.7	83.8
17	**0.6**	*73.7*	74.8	56.8	82.0
	0.7	*75.2*	76.3	58.0	83.5
	0.75	*75.3*	76.4	58.0	83.6
	0.8	*75.3*	76.3	58.1	83.8
	0.85	*75.3*	76.5	57.9	83.6
	0.9	*75.2*	76.2	57.9	83.7
	0.95	*75.2*	75.9	57.8	83.9
	1.0	*74.9*	75.3	57.9	83.8

ity of our HNNP can be maximized. As introduced in the previous section, this stage was implemented using an MLP network. The network was trained with the fundamental back-propagation technique—gradient descent. This training algorithm is relatively simple and straightforward and is the most traditional way for the training of an MLP network. On the other hand, there are other training algorithms which are superior and more efficient, such as the conjugate gradient (CoG) method [90][242], the Levenberge–Marqudt (LM) method [191] etc. It might seem unreasonable for us to select a 'naive' rather than a 'smart' algorithm for our training. Let us explain the rationale beneath.

The LM method is a non-linear optimization algorithm which combines the advantages of the steepest-descent and the Gauss–Newton methods: fast convergence with the steepest-descent method when far from the minimum and fast convergence with the Gauss–Newton method when close to the minimum. The operation of the algorithm is based on

$$H = J^{\mathrm{T}} J, \tag{8.28}$$

Table 8.4. Prediction accuracies of the sequence–structure prediction with various window sizes N and purity levels ρ: $N \in \{19, 21, 23\}$.

Window size (N)	**Purity level (ρ)**	**Accuracies** $Q_3(\%)$	$Q_\alpha(\%)$	$Q_\beta(\%)$	$Q_c(\%)$
19	**0.6**	*73.8*	74.8	57.0	82.0
	0.7	*75.2*	76.1	57.9	83.8
	0.75	*75.4*	76.5	58.2	83.7
	0.8	*75.3*	76.4	58.6	83.5
	0.85	*75.4*	76.5	58.1	83.6
	0.9	*75.3*	76.5	57.9	83.6
	0.95	*75.3*	75.8	58.4	83.9
	1.0	*75.0*	76.1	57.7	83.4
21	**0.6**	*73.7*	74.7	56.7	81.8
	0.7	*75.4*	76.4	58.6	83.5
	0.75	*75.3*	76.3	58.1	83.7
	0.8	*75.5*	76.7	58.5	83.5
	0.85	*75.6*	76.6	58.3	83.9
	0.9	*75.3*	76.3	57.4	84.0
	0.95	*75.1*	75.8	58.5	83.5
	1.0	*74.9*	75.8	58.2	83.1
23	**0.6**	*73.8*	74.6	56.7	82.3
	0.7	*75.3*	76.1	58.1	83.9
	0.75	*75.3*	76.6	57.8	83.6
	0.8	*75.4*	76.8	58.4	83.4
	0.85	*75.5*	76.6	58.3	83.8
	0.9	*75.4*	76.5	57.9	83.7
	0.95	*75.2*	75.9	58.3	83.7
	1.0	*74.9*	76.0	57.9	83.2

where J is the Jacobian matrix that contains first derivatives of the network errors with respect to the weights and biases. In other words, this matrix has a size: number-of-weights-bias $\times$ number-of-patterns. In our problem, the number of training patterns is around 2×10^5. If we were to need 1000 weights (which is quite moderate to our problem) to train these patterns, the memory required to store the Jacobian matrix would be $2 \times 10^5 \times 10^3 \times 8$ bytes (double precision) ≈ 1.6 GB. Such a demand for memory space is beyond the computation resources available to us, not to mention the even more space-demanding inner product between the Jacobian matrix and its transpose. Hence, it is not feasible for us to use the LM method despite its faster convergence.

The back-propagation algorithm searches for the minimum of the error surface in the negative direction of the gradient. The CoG method, however, applies a line search technique for which the search is guided by the conjugate directions. The algorithm is especially suitable to an error minimization prob-

Table 8.5. Prediction accuracies of the sequence-structure prediction with various window size N and purity level ρ: $N \in \{25, 27\}$

Window size (N)	**Purity level (ρ)**	**Accuracies** $Q_3(\%)$	$Q_\alpha(\%)$	$Q_\beta(\%)$	$Q_c(\%)$
25	**0.6**	*73.6*	74.6	56.3	82.1
	0.7	*75.2*	76.0	58.2	83.6
	0.75	*75.3*	76.6	58.0	83.5
	0.8	*75.2*	76.2	57.9	83.5
	0.85	*75.4*	76.5	58.6	83.4
	0.9	*75.4*	76.3	58.3	83.6
	0.95	*75.1*	75.8	58.1	83.6
	1.0	*74.7*	75.9	57.3	83.0
27	**0.6**	*73.4*	73.8	56.3	82.1
	0.7	*75.2*	76.4	58.1	83.3
	0.75	*75.2*	76.2	58.0	83.6
	0.8	*75.2*	76.3	58.0	83.5
	0.85	*75.2*	76.3	57.8	83.5
	0.9	*75.1*	76.2	57.7	83.6
	0.95	*75.1*	76.0	57.9	83.5
	1.0	*74.8*	75.8	57.4	83.2

Table 8.6. Peak accuracies reached at each window size N: a summary

Window size (N)	11	13	15	17	19	21	23	25	27
Q_3	*74.8*	*75.1*	*75.3*	*75.3*	*75.4*	*75.6*	*75.4*	*75.4*	*75.2*

lem with a quadratic error surface. In comparison to the back-propagation method, the CoG method is generally faster in convergence and may help us to escape from some local minima like saddle points. On the other hand, the CoG method has to adopt the batch learning technique such that the direction found is suitable for all training data: the newly found directions have to be tested on all training data against the sum of the squared errors; the method tries several times of searching until an appropriate direction has been found; otherwise, a randomization of the search would be made and the newly found direction is again evaluated for correctness, and a similar scenario repeats until the minimum of the error surface is reached. Each trial of search requires several goes (to check whether the sum of the squared errors is reduced) through all training samples. Hence, the time consumed in the direction search may become significant when the data set used is extremely large in scale and high in dimensionality. As a result, the overall convergence speed might be lower despite the fact that the CoG method requires fewer steps to reach the minimum error. And, our experiments using the CoG method seem to match the above rationale. From our experiments, we have noticed that the method spent most of its time in searching for a new direction that minimized the error. As an example for the comparison of the convergence, the back-propagation algorithm reached its minimum with a sum of the squared errors of 0.23 using around one night of time, while, the CoG could only reach a sum of the squared errors of 0.3 within the same amount of time. As a result, we believe that the back-propagation algorithm is more suitable for our problem.

Strategy

Our second stage predictor is a simple yet large MLP network. The network is simple as it is implemented with a standard three-layer feedforward network and is trained using the fundamental back-propagation technique; the network is also large in that several thousands of neurons are used in the hidden layer and a large amount of data is used for the training.

To set up the network, an online learning method was adopted. That is, the weights of the MLP neural network were updated using the back-propagation technique at each time when an input pattern is present. When all of the input patterns have been used, this is called one *epoch* of the training. At each epoch, the training accuracy and the generalization performance of the network are examined. The training continues if both the training accuracy and the generalization performance are increasing. The training stops at the optimal state where there was no further improvement of generalization with the increase of training efficiency. To obtain the optimal performance, we also need to inspect the predictions with different window size N. The strategy adopted for the experiment is as follows:

1. Randomly start with a window size of 17. Set up the MLP network with the output of the training data from the first stage.
2. Feed the testing data set into the network to carry out the prediction. Record the prediction accuracy.
3. Increase the window size by two. Train the network and carry out the prediction. Compute the prediction accuracy.
4. Repeat step 3 until a significant decrease in prediction accuracy is found.
5. Set the window size to 15. Train the network and carry out the prediction. Compute the prediction accuracy.
6. Decrease the window size by 2. Train the network and carry out the prediction. Compute the prediction accuracy.
7. Repeat step 6 until a significant decrease in the prediction accuracy is found.
8. Finalize the ultimate secondary structure assignment on the testing data set by selecting the best prediction.

Results

Table 8.7 shows the experimental results for window size N varying from 11 to 25. From the table, the best accuracy Q_3 of the T2T prediction is 76.8% when

Table 8.7. Prediction accuracies of the structure–structure prediction for window size $N = \{11, 13, 15, 17, 19, 21, 23, 25\}$.

Window size (N)	**Accuracies**			
	$Q_3(\%)$	$Q_\alpha(\%)$	$Q_\beta(\%)$	$Q_c(\%)$
11	*76.4*	79.6	61.9	81.6
13	*76.7*	79.1	61.9	82.7
15	*76.7*	79.4	61.7	82.5
17	*76.8*	78.9	65.6	81.2
19	*76.6*	79.8	61.8	81.9
21	*76.5*	80.0	61.0	82.1
23	*76.5*	79.8	60.9	82.1
25	*76.4*	80.4	61.2	81.3

the window size N is 17. This is also the best final prediction accuracy that our HNNP could reach. For the best prediction accuracy obtained, HNNP reached the corresponding class accuracy Q_α=78.9% for α-helix, Q_β=65.6% for β-strand and Q_c=81.2% for coil.

Table 8.8 shows the peak accuracies reached in the Q2T and the T2T predictions. From this table, we can clearly see the contributions of the

Table 8.8. Contributions of the sequence–structure and the structure–structure predictions to the final prediction accuracy.

Prediction stage	**Accuracies**			
	$Q_3(\%)$	$Q_\alpha(\%)$	$Q_\beta(\%)$	$Q_c(\%)$
Q2T	*75.6*	76.6	58.3	83.9
T2T	*76.8*	78.9	65.6	81.2

sequence–structure prediction and the structure–structure prediction to the final prediction accuracy. The Q2T prediction, i.e., the RBF neural network, is the crucial part of the PSSP problem and gives rise to a Q_3 accuracy of 75.6%. Our secondary structure prediction almost entirely relied on this stage of prediction. Based on this prediction, the T2T prediction, i.e., the MLP neural network, improves the raw prediction by 1.2%. Although this contribution is minor in comparison to that of the RBF neural network, it is not negligible. The improvement is a 'reward' received from studying the correlations hidden in the protein secondary structures and the improvement is worth treasuring.

From the table, we can also see the change of prediction accuracies for each secondary structure class. The use of a structure–structure prediction improved the prediction of the α-helix by 2.3% and the β-strand by 7.3%. Obviously, the improvements were gained from correcting wrong predictions to the coil state. On the other hand, the correction has also incorrectly predicted some coil states to helix or strand states and this resulted in the drop of Q_c. Despite that, the correction by T2T is still successful as the overall accuracy Q_3 has been improved.

By comparing the prediction accuracies for different classes, we also see that the prediction accuracies for helix, strand, and coil are quite different. This difference, however, is a reflection of the class distribution of the training data. And, this is relevant to the statistical analysis of the class distributions of the training data. The analysis shows that the population ratio for helix, strand, and coil classes in the training is roughly 3:2:4. The most populated coil class achieves the highest prediction accuracy and the least populated strand class achieves the lowest. The reason for such discrimination is due to the discrimination in training. The coil class tends to 'teach' the learning machine more frequently than the strand and helix classes. Hence, the machine becomes more sensitive to the coil class and less sensitive to the other two classes. As a result, such discrimination is reflected in the prediction accuracies of the three classes.

Table 8.9 gives a more thorough evaluation of our final prediction with additional information like segment overlap measure (SOV) and the Matthews' correlations (ρ_α, ρ_β, ρ_c). The result is also compared with one of the existing leading methods PSIPRED [164].

Table 8.9. More thorough performance evaluation of our HNNP and its comparison with the PSIPRED method.

Method	Per-Residue				Matthews			Segment
	$Q_3(\%)$	$Q_\alpha(\%)$	$Q_\beta(\%)$	$Q_c(\%)$	ρ_α	ρ_β	ρ_c	$SOV(\%)$
HNNP	*76.8*	78.9	65.6	81.2	0.72	0.61	0.59	*73.6*
PSIPRED	*76.5*	–	–	–	–	–	–	*73.5*

From the table, we see that our HNNP outperforms the PSIPRED method in both the per-residue accuracy and the segment overlap measure, i.e., 76.8% VS. 76.5% for Q_3 and 73.6% VS. 73.5% for SOV. Although the improvement is not so significant, it at least shows that the HNNP system is feasible for the PSSP problem. It is the first time that the RBF neural network was applied to the PSSP problem. To train the RBF neural network, a distance measure scheme WINDist which tried to address the importance of neighboring residues was proposed. The overlapped clustering technique was applied to enhance the noise rejection capability of the network. All the efforts made to pursue the accurate secondary structure prediction have been rewarding, as the hybrid networking of the RBF neural network with the MLP neural network has reached a comparable performance with the existing leading method.

9

Support Vector Machines for Prediction

In this chapter, we use support vector machines (SVMs) to deal with two bioinformatics problems, i.e., cancer diagnosis based on gene expression data and protein secondary structure prediction (PSSP) [57][58]. For the problem of cancer diagnosis, the SVMs that we use achieved highly accurate results with fewer genes compared to previously proposed approaches. For the problem of PSSP, the SVMs achieved results comparable to those obtained by other methods.

9.1 Multi-class SVM Classifiers

There are often more than two classes in a data set. Therefore, binary SVMs are usually not enough to solve the whole problem. To solve multi-class classification problems, we should divide the whole problem into a number of binary classification problems. Usually, there are two approaches [176]. One is the 'one against all' scheme and the other is the 'one against one' scheme.

In the 'one against all' scheme, if there are N classes in the entire data set, then N independent binary classifiers are built. Each binary classifier is in charge of picking out one specific class from all the other classes. For one specific pattern, all the N classifiers are used to make a prediction. The pattern is categorized to the class that receives the strongest prediction. The prediction strength is measured by the result of the decision function. In the 'one against one' scheme, there must be one (and only one) classifier taking charge of the classification between any two classes. Therefore, for a data set with S classes, $S(S-1)/2$ binary classifiers are used. To get the ultimate result, a voting scheme is used. For every input vector, all the classifiers give their votes so there will be $S(S-1)/2$ votes; when all the classification (voting) is finished, the vector is designated to the class getting the highest number of votes. If one vector gets highest votes for more than one class, it is randomly designated to one of them. In this chapter, we use 'one against one' policy.

9.2 SVMs for Cancer Type Prediction

In recent years, gene-expression-based cancer classifiers have achieved good results in classifying lymphoma [5], leukemia [121], breast cancer [202], liver cancer [51], and so on.

Gene-expression-based cancer classification is challenging due to the following two properties of gene expression data. Firstly, gene expression data are usually very high dimensional. The dimensionality usually ranges from several thousand to over ten thousand. Secondly, gene expression data sets usually contain relatively small numbers of samples, e.g., a few tens. If we treat this pattern recognition problem with supervised machine learning approaches, we need to deal with the shortage of training samples and high-dimensional input features.

Recent approaches to this problem include artificial neural networks [173], an evolutionary algorithm [75], nearest shrunken centroids [307], and a graphical method [38]. Here, we use SVMs to solve this problem.

9.2.1 Gene Expression Data Sets

Three data sets are used in this chapter. One is the small round blue cell tumors (SRBCTs) data set [173]. Another is the lymphoma data set [5]. The last one is the leukemia data set [121]. The details of the data sets are shown in Chap. 3.

We followed the normalization procedure used in [83]. Three steps were taken, i.e., (a) setting a threshold with a floor of 100 and a ceiling of 16000, that is, if a value is greater (smaller) than the ceiling (floor), this value is replaced by the ceiling (floor); (b) filtering, leaving out the genes with $\max/\min \leq 5$ or $(\max - \min) \leq 500$ (max and min refer to the maximum and minimum of the expression values of a gene, respectively); (c) carrying out logarithmic transformation with 10 as the base to all the expression values. 3571 genes survived after these three steps. Furthermore, the data were standardized across experiments, i.e., subtracted by the mean and divided by the standard deviation of each experiment.

9.2.2 A T-test-Based Gene Selection Approach

The t-test is a statistical method proposed by Welch [343] to measure how large the difference is between the distributions of two groups of samples. If a gene shows large distinctions between two groups, the gene is important for classification of the two groups. To find the genes that contribute most to classification, the t-test has been used in gene selection [320] in recent years.

Selecting important genes using the t-test involves several steps. In the first step, a score based on the t-test (named t-score or TS) is calculated for each gene. In the second step, all the genes are rearranged according to their

TSs. The gene with the largest TS is put in the first place of the ranking list, followed by the gene with the second largest TS, and so on.

Finally, only some top genes in the list are used for classification. The standard t-test is applicable to measure the difference between only two groups. Therefore, when the number of classes is more than two, we need to modify the standard t-test. In this case, we use the t-test to measure the difference between one specific class and the centroid of all the classes. Hence, the definition of the TS for gene i can be described as follows:

$$TS_i = \max\{|\frac{\overline{x}_{ik} - \overline{x}_i}{m_k s_i}|, k = 1, 2, ..., K\}, \tag{9.1}$$

$$\overline{x}_{ik} = \sum_{j \in C_k} \overline{x}_{ij} / n_k, \tag{9.2}$$

$$\overline{x}_i = \sum_{j=1}^{n} x_{ij} / n, \tag{9.3}$$

$$s_i^2 = \frac{1}{n - K} \sum_k \sum_{j \in C_k} (x_{ij} - \overline{x}_{ik})^2, \tag{9.4}$$

$$m_k = \sqrt{1/n_k + 1/n}, \tag{9.5}$$

There are K classes. $\max\{y_k, k = 1, 2, ..., K\}$ is the maximum of all y_k. C_k refers to class k that includes n_k samples. x_{ij} is the expression value of gene i in sample j. $\overline{x}_{ik}$ is the mean expression value in class k for gene i. n is the total number of samples. $\overline{x}_i$ is the general mean expression value for gene i. s_i is the pooled within-class standard deviation for gene i.

9.3 Experimental Results

We applied the above gene selection approach and SVMs to process the SRBCT, the lymphoma, and the leukemia data sets.

9.3.1 Results for the SRBCT Data Set

In the SRBCT data set, we firstly ranked the importance of all the genes with TSs. We picked out 60 of the genes with the largest TSs to perform classification. The top 30 genes are listed in Table 9.1. We input these genes one by one to the SVM classifier according to their ranks. That is, we first input the gene ranked no. 1 in Table 9.1. Then, we trained the SVM classifier with the training data and tested the SVM classifier with the testing data. After that, we repeated the whole process with the top two genes in Table 9.1, and then the top three genes, and so on. Figure 9.1 shows the training and the testing accuracies with respect to the number of genes used.

Table 9.1. The 30 top genes selected by the t-test in the SRBCT data set.

Rank	Gene ID	Gene Description
1	810057	Cold shock domain protein A
2	784224	Fibroblast growth factor receptor 4
3	296448	Insulin-like growth factor 2 (somatomedin A)
4	770394	Fc fragment of IgG, receptor, transporter, alpha
5	207274	Human DNA for insulin-like growth factor II (IGF-2); exon 7 and additional ORF
6	244618	ESTs
7	234468	ESTs
8	325182	Cadherin 2, N-cadherin (neuronal)
9	212542	Homo sapiens mRNA; cDNA DKFZp586J2118 (from clone DKFZp586J2118)
10	377461	Caveolin 1, caveolae protein, 22kD
11	41591	Meningioma (disrupted in balanced translocation) 1
12	898073	Transmembrane protein
13	796258	Sarcoglycan, alpha (50kD dystrophin-associated glycoprotein)
14	204545	ESTs
15	563673	Antiquitin 1
16	44563	Growth associated protein 43
17	866702	Protein tyrosine phosphatase, non-receptor type 13 (APO-1/CD95 (Fas)-associated phosphatase)
18	21652	Catenin (cadherin-associated protein), alpha 1 (102kD)
19	814260	Follicular lymphoma variant translocation 1
20	298062	troponin T2, cardiac
21	629896	Microtubule-associated protein 1B
22	43733	glycogenin 2
23	504791	Glutathione S-transferase A4
24	365826	Growth arrest-specific 1
25	1409509	troponin T1, skeletal, slow
26	1456900	Nil
27	1435003	Tumor necrosis factor, alpha-induced protein 6
28	308231	Homo sapiens incomplete cDNA for a mutated allele of a myosin class I, myh-1c
29	241412	E74-like factor 1 (ets domain transcription factor)
30	1435862	Antigen identified by monoclonal antibodies 12E7, F21, and O13

In this data set, we used SVMs with RBF kernels. C and γ were set as 80 and 0.005, respectively. This classifier obtained 100% training accuracy and 100% testing accuracy using the top seven genes. In fact, the values of C and γ have great impact on the classification accuracy. Figure 9.2 shows the classification results with different values of γ. We also applied SVMs with linear kernels (with kernel function $K(\mathbf{X}, \mathbf{X}_i) = \mathbf{X}^{\mathrm{T}}\mathbf{X}_i$) and SVMs with polynomial kernels (with kernel function $K(\mathbf{X}, \mathbf{X}_i) = (\mathbf{X}^{\mathrm{T}}\mathbf{X}_i + 1)^p$ and order $p = 2$) to the SRBCT data set. The results are shown in Fig. 9.3 and Fig.

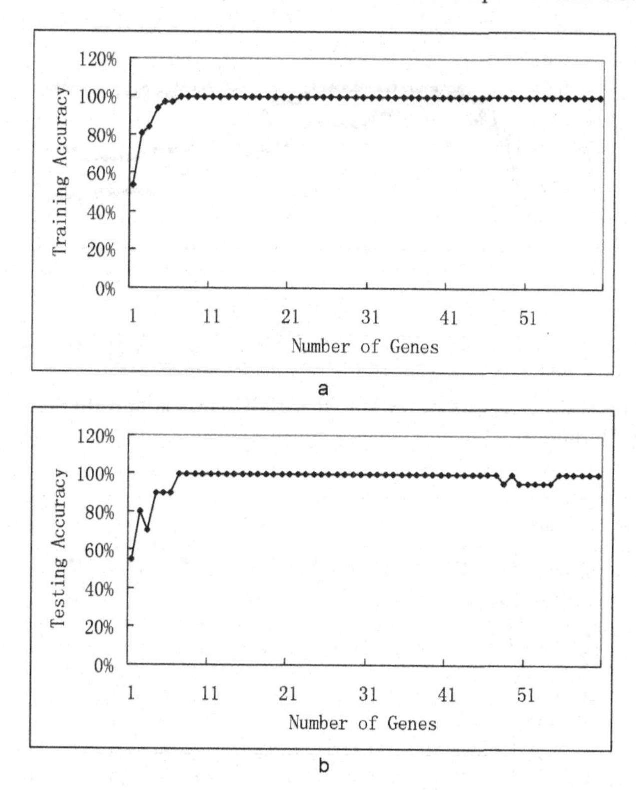

Fig. 9.1. The classification results versus the number of genes used for the SRBCT data set: (a) the training accuracy; (b) the testing accuracy.

9.4. The SVMs with linear kernels and the SVMs with polynomial kernels obtained 100% accuracy with seven and six genes, respectively. The similarity of these results indicates that the SRBCT data set is separable for all the three kinds of SVMs.

For the SRBCT data set, Khan *et al.* [173] 100% accurately classified the four types of cancers with a linear artificial neural network by using 96 genes. Their results and our results of the linear SVMs both proved that the classes in the SRBCT data set are linearly separable. In 2002, Tibshirani *et al.* [307] also correctly classified the SRBCT data set with 43 genes by using a method named nearest shrunken centroids. Deutsch [75] further reduced the number of genes required for reliable classification to 12 with an evolutionary algorithm. Compared with these previous results, the SVMs that we used can achieve 100% accuracy with only six genes (for the polynomial kernel function version, $p = 2$) or seven genes (for the linear and the RBF kernel function versions). Table 9.2 summarizes this comparison.

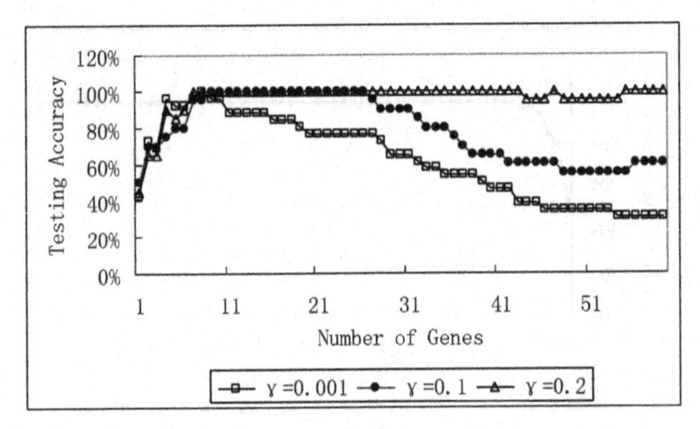

Fig. 9.2. The testing results of SVMs with RBF kernels and different values of γ for the SRBCT data.

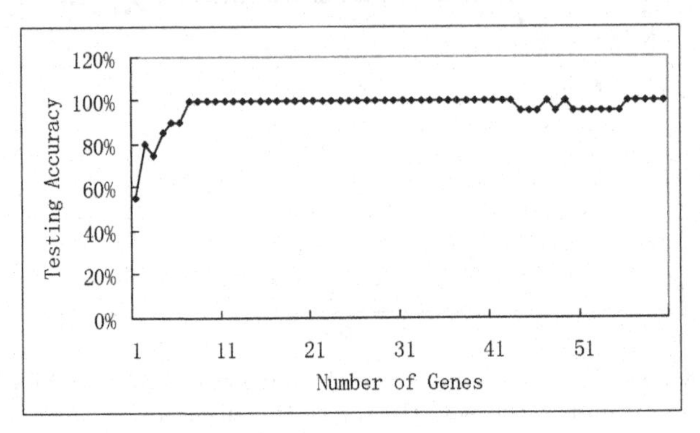

Fig. 9.3. The testing results of the SVMs with linear kernels for the SRBCT data.

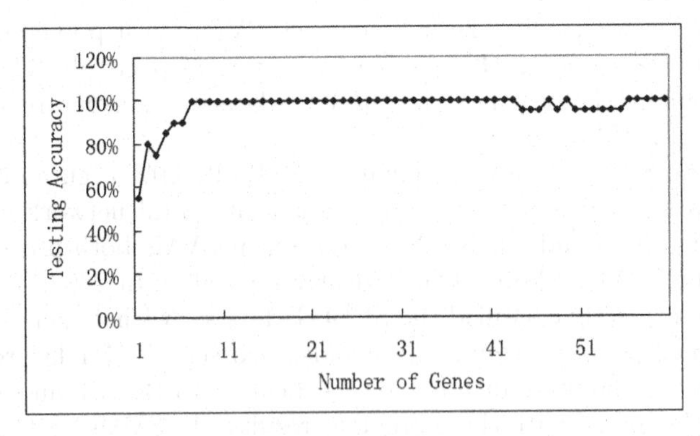

Fig. 9.4. The testing results of the SVMs with polynomial kernels (p=2) for the SRBCT data.

Table 9.2. Comparison of the numbers of genes required by different methods to achieve 100% classification accuracy.

Method	Number of genes required
Linear MLP neural network [173]	96
Nearest shrunken centroids [307]	43
Evolutionary algorithm [75]	12
SVM (linear or RBF kernel function)	7
SVM (polynomial kernel function, $p = 2$)	6

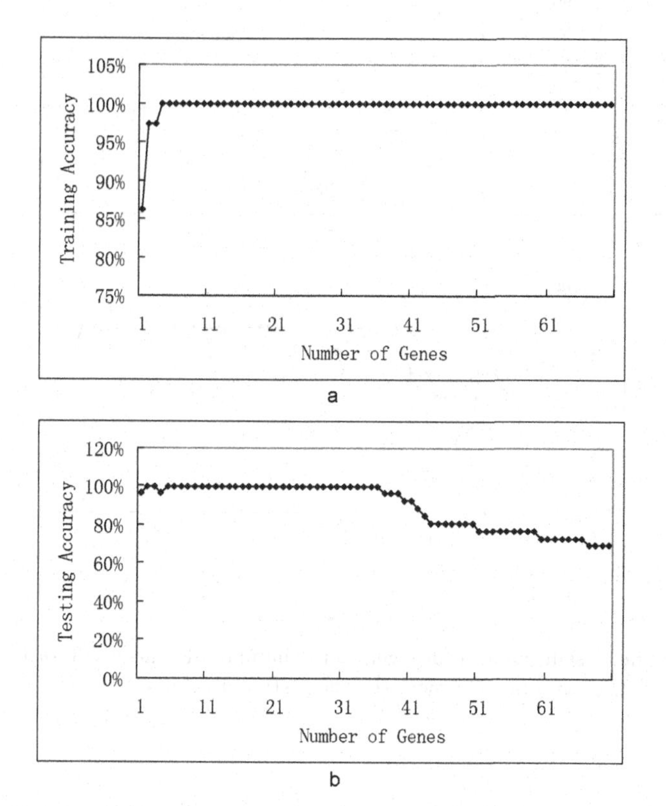

Fig. 9.5. The classification results versus the number of genes used for the lymphoma data set: (a) the training accuracy; (b) the testing accuracy.

9.3.2 Results for the Lymphoma Data Set

In the lymphoma data set, we selected the top 70 genes. The training and testing accuracies with the 70 top genes are shown in Fig. 9.5. The classifiers used here are also SVMs with RBF kernels. The best C and γ obtained are equal to 20 and 0.1, respectively. The SVMs obtained 100% accuracy for both the training and the testing data with only five genes.

For the lymphoma data set, nearest shrunken centroids [308] used 48 genes to give 100% accurate classification. In comparison with this, the SVMs that we used greatly reduced the number of genes required.

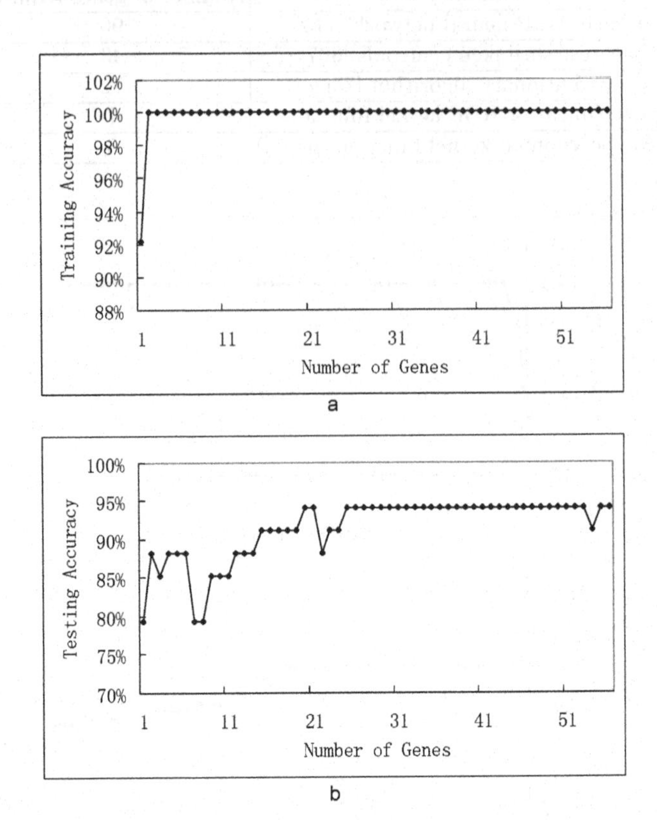

Fig. 9.6. The classification results versus the number of genes used for the leukemia data set: (a) the training accuracy; (b) the testing accuracy.

Results for the Leukemia Data Set

Alizadeh *et al.* [5] built a 50-gene classifier that made one error in the 34 testing samples and, in addition, it could not give strong prediction to another three samples. Nearest shrunken centroids made two errors among the 34 testing samples with 21 genes [307]. As shown in Fig. 9.6, we used the SVMs with RBF kernels with two errors for the testing data but with only 20 genes.

9.4 SVMs for Protein Secondary Structure Prediction

In this section, we use SVMs to solve the PSSP problem. The background information about PSSP has been introduced previously.

The data set used here was originally developed and used by Jones [164]. This data set can be obtained from the website (http://bioinf.cs.ucl.ac.uk/psipred/). The data set contains a total of 2235 protein sequences for training and 187 sequences for testing. All the sequences in this data set have been processed by the online alignment searching tool PSI-Blast (http://www.ncbi.nlm.nih.gov/BLAST/).

As mentioned above, we will conduct PSSP in two stages, i.e., Q2T prediction and T2T prediction.

9.4.1 Q2T prediction

Parameter Tuning Strategy

For PSSP, there are three parameters, i.e., the window size N and SVM parameters (C, γ), to be tuned. N determines the span of the sliding window, i.e., how many neighbors are to be included in the window. Here, we test four different values for N, i.e., 11, 13, 15, and 17.

Searching for the optimal (C, γ) pair is also difficult because the data set used here is extremely large. In [196], an optimal pair was found, $(C, \gamma) = (2, 0.125)$, for the PSSP problem with a much smaller data set (about 10 times smaller compared to the data set used here). Despite the difference of data sizes, we find that their optimal pair also benefits our search as a proper starting point. During our search, we change only one parameter at a time. If the change (increase/decrease) leads to a higher accuracy, we continue to perform a similar change (increase/decrease) next time; otherwise, we reverse the change (decrease/increase). Both C and γ are tuned with this scheme.

Results

Tables 9.3, 9.4, 9.5, and 9.6 show the experimental results for various (C, γ) pairs with the window size $N \in \{11, 13, 15, 17\}$, respectively. Here, Q_3 stands for the overall accuracy; Q_α, Q_β, and Q_c are the accuracies for α-helix, β-strand, and coil, respectively.

From these tables, we can see that the optimal (C, γ) values for window size $N \in \{11, 13, 15, 17\}$ are $(1.5, 0.03)$, $(2, 0.045)$, $(2, 0.04)$, and $(2, 0.03)$, respectively. The corresponding Q_3 accuracies achieved are 73.9%, 74.2%, 74.2%, and 74.1%, respectively. A window size of 13 or 15 seems to be the optimal window size that could most efficiently capture the information hidden in the neighboring residues. The best accuracy achieved is 74.2%, with $N = 13$ and $(C, \gamma) = (2, 0.045)$, or $N = 15$ and $(C, \gamma) = (2, 0.04)$.

Table 9.3. Q2T prediction accuracies of the SVMs with different (C, γ) values: window size $N = 11$.

C	γ	Accuracy			
		$Q_3(\%)$	$Q_\alpha(\%)$	$Q_\beta(\%)$	$Q_c(\%)$
1	0.02	*73.8*	71.7	54.0	85.5
1	0.04	*73.8*	72.4	53.9	85.1
1.5	0.03	*73.9*	72.6	54.2	84.9
2	0.04	*73.7*	73.1	54.4	84.0
2	0.045	*73.7*	73.3	54.5	83.8
2.5	0.04	*73.6*	73.3	54.8	83.4
2.5	0.045	*73.7*	73.3	55.2	83.4
4	0.04	*73.3*	73.4	55.9	82.0

Table 9.4. Q2T prediction accuracies of the SVMs with different (C, γ) values: window size $N = 13$.

C	γ	Accuracy			
		$Q_3(\%)$	$Q_\alpha(\%)$	$Q_\beta(\%)$	$Q_c(\%)$
1	0.02	*73.9*	72.3	54.8	84.9
1.5	0.008	*73.6*	71.4	54.3	85.0
1.5	0.02	*73.9*	72.6	54.7	84.8
1.7	0.04	*74.1*	73.6	54.8	83.4
2	0.025	*74.0*	73.0	55.1	84.3
2	0.04	*74.1*	73.9	55.0	83.9
2	0.045	*74.2*	74.1	55.9	83.5
4	0.04	*73.2*	73.9	55.5	81.7

Table 9.5. Q2T prediction accuracies of the SVMs with different (C, γ) values: window size $N = 15$.

C	γ	Accuracy			
		$Q_3(\%)$	$Q_\alpha(\%)$	$Q_\beta(\%)$	$Q_c(\%)$
2	0.006	*73.4*	70.8	54.2	85.2
2	0.03	*74.1*	73.6	55.6	84.0
2	0.04	*74.2*	73.9	55.7	83.7
2	0.045	*74.0*	73.7	55.4	83.7
2	0.05	*74.0*	73.7	55.4	83.6
2	0.15	*69.0*	63.3	32.7	91.9
2.5	0.02	*74.0*	73.0	55.6	84.0
2.5	0.03	*74.1*	74.0	55.9	83.5
4	0.025	*74.0*	73.8	55.8	83.4

Table 9.6. Q2T prediction accuracies of the SVMs with different (C, γ) values: window size $N = 17$.

C	γ	Accuracy			
		$Q_3(\%)$	$Q_\alpha(\%)$	$Q_\beta(\%)$	$Q_c(\%)$
1	0.125	*70.0*	63.6	36.0	91.3
2	0.03	*74.1*	73.5	56.2	83.7
2.5	0.001	*71.3*	68.1	52.4	83.5
2.5	0.02	*74.0*	68.1	52.4	83.5
2.5	0.04	*74.0*	75.0	55.8	83.1

The original model of SVMs was designed to perform binary classification. To deal with multi-class problems, one usually needs to decompose a large classification problem into a number of binary classification problems. We used the 'one-against-one' scheme [147] in this chapter.

In 2001, Crammer and Singer proposed a direct method to build multi-class SVMs [67]. We also applied such a multi-class SVM to PSSP (http://www.csie.ntu.edu.tw/ cjlin/bsvm/). The results are shown in Table 9.7. Through comparing Table 9.5 and Table 9.7, we found that the multi-class SVMs using Crammer and Singer's scheme [67] and the group of the binary SVMs using 'one-against-one' scheme [147] obtained similar results.

Table 9.7. Q2T prediction accuracies of the multi-class classifier of BSVM with different (C, γ) values: window size $N = 15$.

C	γ	Accuracy			
		$Q_3(\%)$	$Q_\alpha(\%)$	$Q_\beta(\%)$	$Q_c(\%)$
2	0.04	*74.18*	73.90	56.39	84.18
2	0.05	*74.02*	73.68	56.09	83.39
2.5	0.03	*74.20*	73.95	56.85	83.22
2.5	0.035	*74.06*	73.93	56.70	82.99
3.0	0.35	*73.77*	73.88	56.55	82.44

9.4.2 T2T prediction

The T2T prediction uses the output of the Q2T prediction as its input. In T2T prediction, we use the same SVMs as the ones we use in the Q2T prediction. Therefore, we also adopt the same parameter tuning strategy as in the Q2T prediction.

Table 9.8. The T2T prediction accuracies for window sizes $N = 15, 17$, and 19.

Window size (N)	C	γ	Accuracy			
			$Q_3(\%)$	$Q_\alpha(\%)$	$Q_\beta(\%)$	$Q_c(\%)$
15	1	2^{-5}	*72.6*	77.9	60.8	74.3
17	1	2^{-4}	*72.6*	78.0	60.4	74.5
19	1	2^{-6}	*72.8*	78.2	60.1	74.9

Results

Table 9.8 shows the best accuracies reached for window size $N \in \{15, 17, 19\}$ with the corresponding C and γ values. From Table 9.8, it is unexpectedly observed that the structure–structure prediction has actually degraded the prediction performance. A close look at the accuracies for each secondary structure class reveals that the prediction for the coils becomes much less accurate. In comparison to the early results (Tables 9.3, 9.4, 9.5, and 9.6) in the first stage, the Q_c accuracy dropped from 84% to 75%. By sacrificing the accuracy for coils, the predictions for the other two secondary structures improved. However, because coils have a much larger population than the other two kinds of secondary structures, the overall 3-state accuracy Q_3 decreased.

9.5 Summary

For the problem of cancer diagnosis based on microarray data, the SVMs used outperformed most of the previously proposed methods in terms of the number of genes required and the accuracy. Therefore, it is concluded that the SVMs can not only make highly reliable prediction, but also can reduce redundant genes. For the PSSP problem, the SVMs also obtained results comparable with those obtained by other approaches.

10

Rule Extraction from Support Vector Machines

It is a challenging task to obtain explicit knowledge from the solutions of the support vector machine (SVM) for explaining classification decisions. This chapter exploits the fact that the decisions from a non-linear SVM could be decoded into linguistic rules based on the information provided by support vectors and its decision function.

The support vectors of an SVM classifier are sparse representation of the training data. Support vectors are located near the decision boundary. These characteristics of support vectors motivate us to extract rectangular rules based on support vectors and decision functions. Given a support vector of a certain class, cross points between each line, which is extended from the support vector along each axis, and the SVM decision hyper-curve are searched first. A hyper-rectangular rule is derived from these cross points. The hyper-rectangle is tuned by a tuning phase in order to exclude those out-class data points. Finally, redundant rules are merged to produce a compact rule set. Simultaneously, important attributes could be highlighted in the extracted rules. Rule extraction results from our proposed method could follow decisions of SVM classifiers very well. Comparisons between our method and other rule extraction methods are also carried out on several benchmark data sets. Higher rule accuracy is obtained in our method with a fewer number of premises in each rule.

10.1 Introduction

Due to their good generalization performance in solving classification and regression problems, support vector machines (SVMs) [39][40][163] have attracted great interest in recent years. Successful applications of SVMs have been reported in various areas, including but not limited to areas in communication [122], time-series prediction [119], and bioinformatics [34][222]. In many applications, it is desirable to know not only the classification decisions but also what leads to the decisions. However, SVMs offer little insight into the reasons why SVM classifiers have made such final decisions. It is desirable to develop a rule extraction algorithm to reveal knowledge embedded in trained SVMs and represent the classification decisions based on SVM classification results by linguistic rules.

In data mining applications, the task of rule extraction has been widely explored for representing relationship between input attributes and class labels from various classifiers and regression models, such as neural networks (NNs), decision trees, and support vector machines [28][102][111][154][200][285][298]. Though SVMs attracted much attention in various areas in recent years, there lacks robust SVM rule extraction techniques in the literature. Rule extraction from SVMs can facilitate data mining clients in many aspects:

- Increase perceptibility from SVM decisions.
- Refine initial domain knowledge, for example find irrelevant attributes which do not play a role in making decisions.
- Explain hidden data concepts by linguistic rules to clients.
- Find active attributes in data sets.

Nunez *et al.* [231] used clustering to obtain prototype vectors which are centers of clusters. The prototype vectors and support vectors are then used to determine boundaries of rules. The prototype vector of each cluster is used as the center of the ellipsoid that defines a rule. In a cluster region, the support vector which is farthest from the cluster's prototype vector is chosen. The straight line between these two points is used as the first axis of the ellipsoid. Other parameters of the ellipsoid are solved by simple geometry. For the generated ellipsoid, a negative partition test result leads to a rule. The ellipsoid region will be partitioned if its partition test result is positive. The partitioned subregions will be subjected to a partition test again. A similar procedure is followed to generate rules with hyper-rectangular boundaries in [231].

However, according to the algorithm in [231] generating a rule only depends on prototypes and support vectors, which may lead to low rule accuracy because common data points of another class may not be detected and rejected from the rule region by this method. In addition, important information of SVMs provided by decision boundaries are not utilized. And, the determination of boundaries of rules becomes complicated with the increase in data dimensionality because it cannot be solved efficiently by just using a simple geometry. In these clustering approaches, both the number and the accuracy of rules are affected significantly by the choice of the clustering algorithm.

In this chapter, a rule extraction algorithm RulExSVM (rule extraction from support vector machines), first proposed in [108], is described for revealing the relationships between attributes and class labels through linguistic rules. The extracted rules are with hyper-rectangular boundaries and in IF–THEN forms. Each rule corresponds to a support vector and is generated directly based on the relationship between the support vector and the decision function. Given a support vector of a certain class, cross points between lines, along each axis, extended from the support vector and SVM decision hyper-curves are found first. A hyper-rectangular rule is derived from these cross points. Out-class data points which do not have the same class label as the support vector are detected. The hyper-rectangular rule is tuned by a tuning phase in order to exclude those out-class data points. Finally, rules are merged to obtain a more compact rule set.

C in Eq. (1.30) is the regularization constant. Support vector i with $\alpha_i = C$ falls in the region between two separating hyper-curves. There might be support vectors between two separating hyper-curves. It is noted that those support vectors, such as support vectors I and J shown in Fig. 10.1, which lie between two separating hyper-curves, are not used for generating rules because they might not be correctly classified.

Many methods have been proposed to solve the optimization problem [279][326] for obtaining SVM classifiers. Based on the solution of the optimization problem, in the classification phase, a data point $\mathbf{x}$ is classified by computing the sign of the decision function:

$$f(\mathbf{x}) = \sum_{i=1}^{N_s} \alpha_i y_i K(\mathbf{s}_i, \mathbf{x}) + b. \tag{10.1}$$

In order to distinguish support vectors from other data samples, we use $\mathbf{s}_i$ to represent the ith support vector. Actually, $\mathbf{s}_i$ is also a data sample. N_s is the number of support vectors.

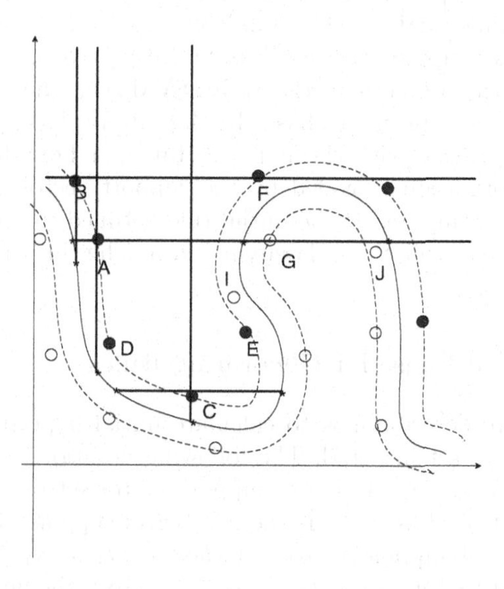

Fig. 10.1. Separating hyper-curves in the original space, support vectors, and the decision function. (© 2005 IEEE) We thank the IEEE for allowing the reproduction of this figure, first appeared in [108].

10.2 Rule Extraction

In this section, the rule extraction procedure will be described. This rule extraction method can extract rules from multi-class data sets. Based on the trained SVM classifier, support vectors of *class 1* are used for generating rules for *class 1*. The rule extraction method could be easily extended to multi-class problems. One-against-all policy is employed for classifying multi-class data sets in SVM training. In the rule extraction algorithm, the current class processed is referred to as *class 1*. All the other classes are referred to as *class 2*.

Support vectors are the skeleton of the training data set. If data points except for support vectors are removed from the training set, the same separating hyper-curves could be obtained in the retrained SVM. Since support vectors can support separating hyper-curves, we start from each support vector to extract rules with hyper-rectangular boundaries.

For illustrating how to obtain initial rules based on the information of an SVM classifier, an example is considered in the two-dimensional space. In Fig. 10.2, black points are the support vectors of *class 1* and white points are the support vectors of *class 2*. For each axis, a line, parallelling the axis, starting from a support vector of *class 1* can be extended unlimitedly in two directions. The cross points between the line and the decision boundary can be calculated. Take for example, for support vectors A and C, cross points between the extended lines and the decision boundary are shown in Fig. 10.2. Based on these cross points, the initial boundaries of the hyper-rectangular rules can be obtained and are shown as rectangles with dashed lines in Fig. 10.3 (a) and (b) for support vectors A and C respectively.

There are three phases in the RulExSVM, i.e., the initial phase, the tuning phase, and the pruning phase. In the initial phase, given a support vector of *class 1*, a rule with the hyper-rectangular boundary is generated based on the information provided by the support vector and the decision boundary. In the tuning phase, the initial rule is tuned towards the direction improving the rule accuracy for classifying data. The three phases are stated as follows.

10.2.1 The Initial Phase for Generating Rules

In this section, we describe how to calculate initial hyper-rectangular rules for a two-class data set in detail. The following notations are used. n is the dimension of the data set. $A1$ is the support vector set of *class 1*. $A2$ is the support vector set of *class 2*. N_1 is the number of support vectors of *class 1*. N_2 is the number of support vectors of *class 2*. $N_s = N_1 + N_2$ is the total number of support vectors. $\mathbf{s}_m = \{s_{m1}, s_{m2}, ..., s_{mn}\}$ is the mth support vector of *class 1*. $\mathbf{x} = \{x_1, x_2, ..., x_n\}$ is a pattern of data. Note that all attributes of data points are normalized to lie in [0, 1].

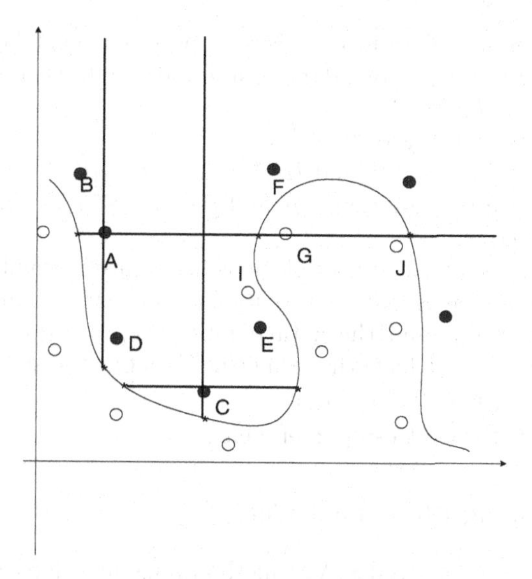

Fig. 10.2. Cross points. (© 2005 IEEE) We thank the IEEE for allowing the reproduction of this figure, first appeared in [108].

The rule with the hyper-rectangular boundary derived from support vector $\mathbf{s}_m$ of *class 1* can be represented by:

$$\{s_{mi} + \lambda_{2i} \geq x_i \geq s_{mi} - \lambda_{1i}, i = 1, ..., n\}, \tag{10.2}$$

where $1 \geq \lambda_{pi} \geq 0, p = \{1, 2\}$.

Let:

$$L_o(i) = s_{mi} - \lambda_{1i} \tag{10.3}$$

and

$$H_o(i) = s_{mi} + \lambda_{2i} \tag{10.4}$$

Here $H_o(i)$ and $L_o(i)$ give the upper limit and the lower limit of the hyper-rectangular rule along the ith dimension, respectively. Based on the decision function $f(\mathbf{x})$ which distinguishes *class 1* from *class 2*, L_o and H_o are initially determined in the procedure of searching rules.

Given the trained SVM, the rule based on support vector $\mathbf{s}_m$ can be generated as follows:

1. Set $l = 1$, (l refers to dimension).
2. Calculate x_l subject to $f(\mathbf{x}) = 0$ and $x_j = s_{mj}$ ($j = 1, ..., n$ and $j \neq l$) by the Newton's method [64][252].
3. Determine L_o and H_o according to the solutions of the problem in step 2. The number of solutions of x_l may be different under different data distributions:

a) If there is no solution, i.e., there is no cross point between the line extended from $\mathbf{s}_m$ along dimension l and the decision boundary, then $L_o(l) = 0$, $H_o(l) = 1$.
b) If there is one solution:
If $s_{ml} \geq x_l$, $L_o(l) = x_l$ and $H_o(l) = 1$, else $L_o(l) = 0$ and $H_o(l) = x_l$.
c) If there are two solutions x_{l1} and x_{l2} $(x_{l1} \leq x_{l2})$: $L_o(l) = x_{l1}$ and $H_o(l) = x_{l2}$.
d) If there are more than two solutions, the nearest neighbors x_{l1} and x_{l2} of s_{mj} are chosen from the solutions under the conditions $x_{l1} \leq s_{mj}$ and $x_{l2} \geq s_{mj}$ and the data points $\{\mathbf{x}|x_j = s_{mj}, j = 1, ..., n, j \neq l, x_{l1} \leq x_l \leq x_{l2}\}$ have the same class label as the support vector $\mathbf{s}_m$: $L_o(l) = x_{l1}$ and $H_o(l) = x_{l2}$.

4. $l = l + 1$, if $l < n$, go to step 2, else end.

10.2.2 The Tuning Phase for Rules

The tuning phase of the RulExSVM method aims at tuning the initial rules for improving rule accuracy by the removal of outliers. Whenever a pattern $\mathbf{x}$ of *class 2* falls into the region of a rule of *class 1*, the rule is tuned to remove the outlier from the region.

In this rule extraction method, rectangular rules are extracted by splitting the data space into rectangles. It is expected to obtain rules covering as large a data space as possible. For the purpose of generating a compact rule set, rules with larger volumes are preferred. In an n-dimensional data space, the outlier is rejected by chopping the rule along a certain dimension in which the volume of the hyper-rectangular rule is larger than those along other dimensions. The detailed steps of RulExSVM are as follows:

1. For a two-class data set, we train a support vector machine first.
2. Choose a support vector from $A1$ which is the support vector set of *class 1* to calculate the initial hyper-rectangular rule region based on the trained support vector machine. Details can be found in Sect. 10.2.1.
3. Search all samples of *class 2* which are included in the rule region. Assume that there are K samples of *class 2* in the rule region. We refer to K samples as a sample subset Q. Randomly choose a sample from Q.
4. Calculate distances from the sample to boundaries of the hyper-rectangular rule along each dimension.
5. Remove the sample from the rectangle by shrinking the hyper-rectangular rule along the axis which can maintain the maximum volume of the hyper-rectangle.
6. Check the samples of *class 2* left in the new rule region; if $K > 0$, randomly choose a sample from Q, go to Step 4, else go to the next step.
7. Remove the support vector from $A1$. If $A1$ is not empty, go to step 2, else end.

10.2.3 The Pruning Phase for Rules

As the final phase of the RulExSVM method, the pruning phase aims at simplifying the rule set by removing redundant rules. Rules which classify data patterns from different classes may be overlapped with each other. If a rule is totally overlapped by another rule, the rule would be considered as a redundant one. Then, a pruning measure is taken to remove redundant rules from the rule set.

In order to remove redundant rules, we do the following: (1) find the patterns falling into each rule region, (2) if the set of patterns in a certain rule region is a subset of patterns covered by another rule, the rule is removed from the rule set, (3) repeat the pruning to remove all redundant rules.

Rule merging is another pruning task, which can also help to simplify the rule set, however, we will not discuss rule merging in this book.

10.3 Illustrative Examples

In this section, we will illustrate the rule extraction method by two data sets. The first data set is a binary-class data set with discrete attributes. The second data set is a multiple-class data set with continuous attributes. For a multi-class data set with M classes, rule extraction is carried out for M binary-class data sets, i.e., one-against-all policy is employed for extracting rules for each class. When training SVM classifiers and extracting rules, we normalized all the attributes to the interval [0, 1]. In the expression of rules, the attributes will be transformed to their original ranges.

10.3.1 Example 1 — Breast Cancer Data Set

This data set have discrete attributes in the interval [1, 10].

RulExSVM extracts rules based on trained SVM classifiers. The parameters of SVM $\{\sigma, C\}$ are determined using 5-fold cross-validation.

To determine a rule, we first choose a support vector $\mathbf{s}_m$ of *class 1* randomly. Newton's method is employed to find the initial rule derived from this support vector. In order to reduce the calculation time of Newton's method when searching for the cross points between lines extending from the selected support vector along each axis and the decision boundary of the support vector machine, 11 values $\{0, 0.1, 0.2, 0.3, 0.4, 0.5, 0.6, 0.7, 0.8, 0.9, 1.0\}$ are taken from the interval [0, 1] along axis l. $\mathbf{x} = \{x_1, x_2, ..., x_n\}$. The values of $f(\mathbf{x})$ are then calculated subject to $\{x_j = s_{mj}\}$ ($j = 1, ..., n$ and $j \neq l$). x_l equals each of the 11 values. Define $\tilde{f}(x_l) := f(\mathbf{x})$. In the 11 results of $\tilde{f}(x_l)$, two neighbors $\{x_1, x_2\}$ whose signs are different are located. Let $x_l = (x_1 + x_2)/2$. If the signs of all of $\tilde{f}(x_l)$'s are the same, x_l equals x, which corresponds to the smallest $\tilde{f}(x)$. This x_l serves as the starting point for Newton's method.

In our algorithm, if $|\tilde{f}(x_l)| \leq 0.001$, then the solutions of $\tilde{f}(x_l) = 0$ are considered found. If there are no cross points (no solutions) along the axis, the rule interval is [0, 1] along the axis.

For the Breast cancer data set, we obtain seven rules. Due to space limitation, we only show the first two rules here:

Rule 1:
IF Attribute 1 $\in \{1,2,3\}$
AND Attribute 3 $\in \{1,2,3,4\}$
AND Attribute 4 $\in \{1,2,3,4\}$
AND Attribute 8 $\in \{1,2,3,4\}$
THEN class label is Benign.
Rule 2:
IF Attribute 1 $\in \{1,2,3,4,5,6\}$
AND Attribute 3 $\in \{1,2,3\}$
AND Attribute 8 $\in \{2,3,4,5,6,7,8,9,10\}$
AND Attribute 9 $\in \{2,3,4,5,6,7,8\}$
THEN class label is Benign.
...
Default rule:
Else class label is Malignant.

In the RulExSVM method, a set of rules is obtained for the breast cancer data set and the rule accuracy for classification is 97.51%. These rules will describe the Benign cases, and the Malignant cases are default. The characteristics of the Malignant cases are considered as the ones opposite to those presented in the rule set above.

Compared with the rule extraction results in [286], in which on average 2.9 rules were generated with accuracy 94.04%, higher rule accuracy is obtained by our method though the number of rules is slightly higher. The rule accuracy is also higher than the result in our previous work based on radial basis function (RBF) neural networks [102][104]. It is also observed in our rule set above that the rule accuracy is obtained without the contribution of some attributes, such as attribute 5. When we use the attributes presented in the rule set as the inputs to an SVM, we can obtain the same classification result as that obtained by using the whole original attributes as inputs to the SVM. Some attributes may be not active in determining class labels. This observation is important especially in medical diagnosis. Thus, doctors can pay more attention to those active attributes for analyzing causes and symptoms of a disease. This point is an advantage of rule decision over SVM decision though the accuracy of black-box SVM classifiers is usually higher than the accuracy of rules.

10.3.2 Example 2 — Iris Data Set

The second data set used for illustrating the rule extraction method is the Iris data set. Three rules are obtained by our RulExSVM. Only attribute 3 and attribute 4 are present in our rule set. Four other rule sets obtained in [28][145][212][231] could be checked in corresponding papers. Compared with the rules extracted from support vector machines [231], higher rule accuracy is obtained in our RulExSVM method with a fewer number of premises in each rule. In [212], rules were extracted for the Iris data set based on radial basis function (RBF) neural networks. The accuracy of the rules in [212] is lower than ours and with all attributes used in the rule set. In [145], three rules were extracted for the Iris data set with 97.33% accuracy and a hyper-plane decision boundary. In the RulExSVM method, the same number of rules and better rule accuracy are obtained, and the rules extracted have a hyper-rectangular decision boundary. The accuracy of our rules is better than the accuracy of the rules extracted from MLP (multi-layer perceptron) neural networks in [28]. In addition, through the rules extracted by RulExSVM, it is observed that it may not affect the concept of the Iris data set with the removal of attributes 1 and 2.

10.4 Experimental Results

Data sets Chess, Mushroom, Iris, Pima-diabete, Breast cancer, and Wine are used in our experiments. All of these data sets can be obtained from the UCI database [223]. The characteristics of the data sets used here are shown in Table 10.1. Discrete and numerical attributes can be found in the data sets.

In Table 10.2, the number of support vectors in SVM classifiers for separating each class from other classes is shown together with the classification accuracy based on trained SVM classifiers. The information of rules extracted are shown in Table 10.3. The number of premises of each rule is calculated on average. In this table, the fidelity shows that rules extracted match the SVM classifier well.

In the experiments, we use only support vectors for generating initial rules and tuning rules by considering training data points. In Table 10.4, time (seconds) consumed for training SVM classifiers and extracting rules is presented. The rule extraction program is written in matlab. The computer has a 2.53 GHz CPU.

We compared our rule extraction algorithm with other rule extraction algorithms. For the Pima diabete data set, we obtained rules with the accuracy of 80.29% which is higher than rules in [145] in which rules with 76.3% accuracy were extracted.

In [286], rules were obtained for the Mushroom data set with two rules, and 98.12% accuracy. For the Mushroom data set, seven rules with 100% accuracy are obtained by our RulExSVM method.

Table 10.1. Characteristics of data sets used (num-attri: numeric attributes, dis-attri: discrete attributes).

Data sets	Patterns	Num-attri	Dis-attri	Classes
Chess	3196	0	36	2
Mush-room	8124	0	22	2
Iris	150	4	0	3
Wine	178	13	0	3
Pima diabete	768	8	0	3
Breast cancer	683	0	9	2

Table 10.2. SVM classification results.

Data sets	SVM accuracy	Number of support vectors
Chess	98.75%	762
Mushroom	100%	250
Iris	97.5%	30
Wine	99.3%	39
Pima diabete	78.83%	207
Breast cancer	97.8%	66

Table 10.3. Comparison of classification results from SVM classifiers and rules.

Data sets	Rule accuracy	Rules	Premises /per rule	Fidelity
Chess	99%	12	2.33	99.42%
Mush-room	100%	7	3.33	100%
Iris	98%	3	1.33	99.19%
Wine	99.3%	6	4.3	99.3%
Pima diabete	80.29%	16	4.46	98.21%
Breast cancer	97.51%	7	5.3	99.27%

10.5 Summary

After an SVM classifier is obtained based on the training data set, the rule-extraction algorithm RulExSVM is implemented in three phases: first, initial rules are determined based on the trained SVM classifier by calculating cross

Table 10.4. SVM training time and time spent on rule extraction.

Data sets	SVM training time (s)	Rule extraction time (s)
Chess	5.43	1419.3
Mushroom	2.56	895.43
Iris	0.56	1.14
Wine	0.32	49.6
Pima diabete	0.46	201.968
Breast cancer	0.11	128.7

points between support vectors and the decision boundary along each axis; second, rules are tuned based on the criterion that excludes data points of other classes from the rule region and keeps the rule region as large as possible; third, the rules which are overlapped completely by other rules are pruned. In this book, we only explore rule extraction from SVM classifiers with non-linear RBF kernel functions. The rule extraction procedure reported here could be easily extended to SVMs with any other type of kernel functions.

It is shown in Table 10.4 that the computational time of rule extraction is far higher than the time for SVM training. For large-scale data, training time of SVM is a big problem which makes on-line processing of large-scale data impractical. A solution to this problem can be parallel computing. In addition, writing the rule extraction program in C language and improving the efficiency of the algorithm are alternative ways to reduce the time spent on rule extraction.

In the rectangular rules extracted by the described algorithm, important attributes are highlighted since unimportant rules are not present or scarce in the rule set. Rule extraction results from the RulExSVM method shows that rules follow SVM decisions very well. Comparisons between our method and other rule extraction methods show that higher rule accuracy is obtained in this method with a fewer number of premises in each rule.

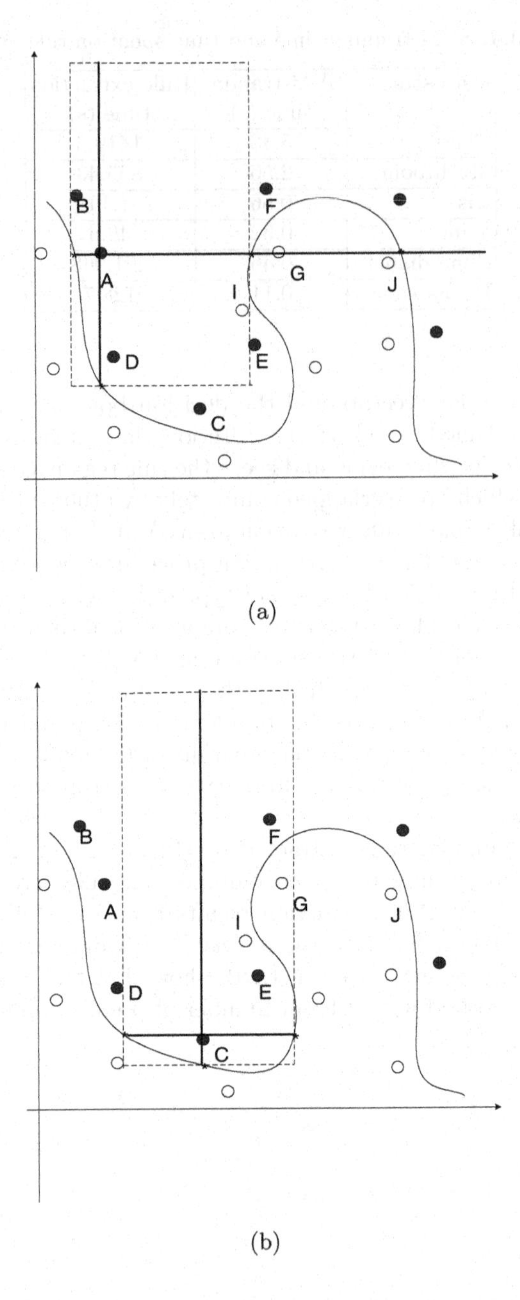

Fig. 10.3. (a) Initial rule generated based on support vector A and (b) initial rule generated based on support vector C. (© 2005 IEEE) We thank the IEEE for allowing the reproduction of this figure, first appeared in [108].

glossary names

ALL	acute lymphoblastic leukemia
AML	acute myeloid leukemia
ANN	artificial neural network
ANN-DT	artificial neural-network decision tree algorithm
AR	autoregression
ART	adaptive resonance theory
BP	back-propagation algorithm
CART	classification and regression tree
CG	compatibility grade
CGA	clustering genetic algorithm
CLL	chronicle lymphocytic leukaemia
COA	centroid of Area
CoG	conjugate Gradient
CSNNs	cost-sensitive neural networks
CSBP	cost-sensitive back-propagation
DLBCL	diffuse large B-cell lymphoma
DDR	data dimensionality reduction
DTB	discrete-time backpropagation neural network
EWS	ewing family of tumors
FAM	fuzzy associative memories
FGA	fuzzy GA
FL	follicular lymphoma
FNNs	fuzzy neural networks
GA	genetic algorithms
HCC	hepatocellular carcinoma
HMM	Hidden Markov Model
HNNP	hybrid Neural Network Predictor
IIR	infinite impulse response
LDA	linear discriminant analysis
LDC	linear discriminant classifier
LGA	local genetic algorithm

LLS	linear least square
LM	levenberge-marqudt
LMS	least mean square
LOG	logistic classifier
LVQ	learning vector quantization
MF	membership function
MIFS	mutual information based feature selection
MLP	multi-layer perceptron
MLD	mean local density
MOM	mean of maximum
MSE	mean squared error
NB	neuroblastoma
NHL	non-hodgkin lymphoma
NMSE	normalized mean squared error
OLS	orthogonal least square learning algorithm
PCA	principal components analysis
PCs	principal components
PDF	probability density function
PSSM	position-specific scoring matrix
PSSP	protein secondary structure prediction
QDC	quadratic discriminant classifier
Q2T	sequence-structure
RBF	radial basis function
RMS	rhabdomyosarcoma
SBP	standard back- propagation
SBS	sequential backward se- lection
SCM	separability and correlation measure
SFS	sequential forward selection
SOV	segment overlap measure
SRBCTs	small round blue cell tumors
SVMs	support vector machines
T2T	structure-structure
TDNN	the time delay neural network
TDNNGF	time delay neural network with global feedback
TSs	t-scores
VIA	validity interval analysis
WD	wavelet decomposition
WP-MLP	wavelet packet MLP
WA	weighted accuracy
WTA	winner-take-All

A

Rules extracted for the Iris data set

Rules extracted by McGarry [212] using RBF networks for the Iris data set with an accuracy of 40% for the test data set:

Rule 1
IF the sepal length is within the interval (4.4, 5.7)
AND the sepal width is within the interval (2.9, 4.4)
AND the petal length is within the interval (1.3, 1.5)
AND the petal width is within the interval (0.2, 0.4)
THEN the class label is Setosa.

Rule 2
IF the sepal length is within the interval (4.9, 6.9)
AND the sepal width is within the interval (2.0, 3.1)
AND the petal length is within the interval (3.5, 5.0)
AND the petal width is within the interval (0.4, 1.0)
THEN the class label is Versicolor

Rule 3
IF the sepal length is within the interval (5.8 , 7.2)
AND the sepal width is within the interval (2.8 , 3.1)
AND the petal length is within the interval (4.5 , 5.8)
AND the petal width is within the interval (1.5, 1.7)
THEN the class label is Virginica

The rules extracted by Bologna *et al.* [28] (A_s is the sepal area and A_p is the petal area):

Rule 1
If $\{A_s - 3.98A_p \geq 2.34\}$
and $\{11.21 \leq A_s - 5.56A_p \leq 21.87 \text{ } or \text{ } A_p - 0.18A_s = 1.47\}$
then the class label is Setosa.

Rule 2
If $\{-0.23 \leq A_p - 0.25A_s \leq 3.45$ *or* $A_p - 0.25A_s = 3.89$ *or* $A_p - 0.25A_s = 4.09\}$
and $\{1.18 \leq A_p - 0.18A_s \leq 5.22$ *or* $A_p - 0.18A_s = 0.83$ *or* $A_p - 0.18A_s = 1.01\}$
then the class label is Versicolor.
Rule 3
If $\{A_p - 0.25A_s \geq 3.85$ *or* $A_p - 0.25A_s = 3.21\}$
and $\{A_p - 0.18A_s \geq 4.48\}$
then the class label is Virginica.

The rules extracted by Hruschka and Ebecken [145] ($a3$: petal length, $a4$: petal width):

Rule 1
If ($a3 < 2.804$) then the class label is Setosa.
Rule 2
If ($a3 > 2.804$) and ($a3 < 4.974$) and ($a4 < 1.678$) then the class label is Versicolor.
Default rule: the class label is Virginica.

References

1. Abrao, P.J., Alves da Silva, A.P., Zambroni de Souza, A.C. (2002): Rule Extraction from Artificial Neural Networks for Voltage Security Analysis. Proceedings of the 2002 International Joint Conference on Neural Networks. 3, 2126–2131
2. Agrawal, R., Imielinski, T., Swami, A. (1993): Database Mining: a Performance Perspective. IEEE Transactions on Knowledge and Data Engineering. 5, 914–925
3. Aha, D.W., Kibler, D., and Albert, M.K. (1991): Instance-based learning algorithms. Machine Learning. 6, 37–66
4. Aldenderfer, M.S., Blashfield, R.K. (1984): Cluster Analysis. Sage Publications, London
5. Alizadeh A.A., et al. (2000): Distinct Types of Diffuse Large B-cell Lymphoma Identified by Gene Expression Profiling. Nature. 403, 503–511
6. Altschul, S.F., Schaffer, A.A., Zhang, J., Zhang, Z., Miller, W., Lipman, F.J. (1997): Gapped BLAST and PSI-BLAST: a New Generation of Protein Database Search Programs. Nucleic Acids Research, 25, 3389–3402
7. Ando, T., Suguro, M., Hanai, T., Kobayashi, T., Honda, H., Seto, M. (2002): Fuzzy Neural Network Applied to Gene Expression Profiling for Predicting the Prognosis of Diffuse Large B-cell Lymphoma. Japanese Journal of Cancer Research. 93, 1207–1212
8. Andre, D., Bennett, F.H., Koza, J.R. (1996): Evolution of Intricate Long-Distance Communication Signals in Cellular Automata Using Genetic Programming. Proceedings of the Fifth International Workshop on the Synthesis and Simulation of Living Systems. Cambridge, MA: MIT Press
9. Andrews, R., Diederich, J., Tickle, A.B. (1995): Survey and Critique of Techniques for Extracting Rules from Trained Artificial Neural Networks. Knowledge-Based Systems. 8, 373–389
10. Baeck, T., Fogel, D.B., Michalewicz, Z. (eds.) (1997): Handbook on Evolutionary Computation. Oxford University Press, New York, and Institute of Physics Publishing, Bristol
11. Baggenstoss, P.M. (1998): Class-Specific Feature Sets in Classification. IEEE International Symposium on Intelligent Control (ISIC). 413–416
12. Baggenstoss, P.M. (1999): Class-Specific Feature Sets in Classification. IEEE Transactions on Signal Processing. 47, 3428–3432
13. Baggenstoss, P.M. (2004): Class-Specific Classifier: Avoiding the Curse of Dimensionality. IEEE Aerospace and Electronic Systems Magazine, 19, 37–52

14. Bailey, A. (2001): Class-Dependent Features and Multicategory Classification. Ph.D. Thesis. University of Southampton
15. Bakshi, B.R., Stephanopoulos, G. (1992): Wavelets as Basis for Localized Learning in a Multi-Resolution Hierarchy. International Joint Conference on Neural Networks. 2, 140–145
16. Baldi, P., Brunak, S., Frasconi, P., Soda, G. and Pollastri, G. (1999): Exploiting the past and the future in protein secondary structure prediction. Bioinformatics. 15(11), 937–946
17. Baraldi, A., Blonda, P., Satalino, G., D'Addabbo, A., Tarantino, C. (2000): RBF Two-Stage Learning Networks Exploiting Supervised Data in the Selection of Hidden Unit Parameters: An Application to SAR Data Classification. IEEE 2000 International Geoscience and Remote Sensing Symposium. 2, 672–674
18. Battiti, R. (1994): Using Mutual Information for Selecting Features in Supervised Neural Net Learning. IEEE Transactions on Neural Networks. 54, 537–550
19. Berardi, V.L., Zhang, G.P. (1999): The Effect of Misclassification Costs on Neural Network Classifiers. Decision Sciences. 30(3), 659–682
20. Berry, M.J.A., Gordon, S.L. (2000): Mastering Data Mining: The Art and Science of Customer Relationship Management. John Wiley & Sons, Inc. New Jersey
21. Bins, J., Draper, B.A. (2001): Feature Selection from Huge Feature Sets. Eighth IEEE International Conference on Computer Vision. 2, 159–165
22. Bishop, C.M. (1995): Neural Network for Pattern Recognition. Oxford University Press Inc., New York
23. Blake, C., Keogh, E., and Merz, C.J. (1998): UCI Repository of machine learning databases [http://www.ics.uci.edu/ ~mlearn/MLRepository.html], University of California, Department of Information and Computer Science: Irvine, CA
24. http://www.ncbi.nlm.nih.gov/BLAST/
25. Bohanec, M., Rajkovic, V. (1990): Expert system for decision making. Sistemica. 1(1), 145–157
26. Bohr, H., Bohr, J., Brunak, S., Cotterill, R.M.J., Lautrup, B., Nrskov, L., Olsen, O., and Petersen, S. (1988): Protein Secondary Structure and Homology by Neural Networks. FEBS Letters. 241, 223–228
27. Bollacker, K.D., Ghosh, J. (1996): Mutual Information Feature Extractors for Neural Classifiers. IEEE International Conference on Neural Networks. 3, 1528–1533
28. Bologna, G., Pellegrini, C. (1998): Constraining the MLP Power of Expression to Facilitate Symbolic Rule Extraction. Proc. IEEE World Congress on Computational Intelligence. 1, 146–151
29. Boser, B., Guyon, I., Vapnik, V.N. (1992): A Training Algorithm for Optimal Margin Classifiers. Fifth Annual Workshop on Computational Learning Theory. 144–152
30. Breiman, L., Friedman, J.H., Olshen, R.A., Stone, C.J.: Classification and Regression Trees. Belmont, CA: Wadsworth, 1984.
31. Bridle, J.S. (1990): Training stochastic model recognition algorithms as networks can lead to maximum mutual information estimation of parameters. Advances in Neural Information Processing Systems. 2, 211–217
32. Brill, F.Z., Brown, D.E., Martin, W.N. (1992): Fast Generic Selection of Features for Neural Network Classifiers. IEEE Transactions on Neural Networks. 3, 324–328
33. Broomhead, D.S., Lowe, D. (1988): Multivariable Functional Interpolation and Adaptive Networks. Complex Systems. 2, 321–355

34. Brown, M.P.S., Grundy, W.N., Lin, D., Critianini, N., Sungnet, C., Furey, T.S., Ares, M., Haussler, D. (2000): Knowledge-Based Analysis of Microarray Gene Expression Data Using Support Vector Machines. Proceedings of National Academy of Sciences. 97, 262–267
35. Brown, M. and Harris, C.(1994): Neurofuzzy adaptive modelling and control. Prentice-Hall
36. Bruzzone, L., Prieto, D.F. (1999): A Technique for the Selection of Kernel-Function Parameters in RBF Neural Networks for classification of Remote-Sensing Images. IEEE Transactions on Geoscience and Remote Sensing. 37, 1179–1184
37. Buckley, J.J., Hayashi, Y. (1994): Can Fuzzy Neural Nets Approximate Continuous Fuzzy Functions. Fuzzy Sets System. 61, 43–51
38. Bura, E., Pfeiffer, R.M. (2003): Graphical Methods for Class Prediction using Dimension Reduction Techniques on DNA microarray data. Bioinformatics, 19, 1252–1258
39. Burges, C.J.C. (1996): Simplified Support Vector Decision Rules. 13th International Conference on Machine Learning. San Mateo, CA. 71–77
40. Burges C.J.C. (1998): A Tutorial on Support Vector Machines for Pattern Recognition. Data Mining and Knowledge Discovery. 2, 955–974
41. Cai, L.Y., Kwan, H.K. (1998): Fuzzy Classifications Using Fuzzy Inference Networks. IEEE Transactions on Systems, Man, and Cybernetics, Part B: Cybernetics. 28, 334–347
42. Campbell, C. (2002): Kernel Methods: a Survey of Current Techniques. Neurocomputing. 48, 63–84
43. Chakraborty, B., Sawada, Y. (1995): Feature Subset Evaluation Using Fuzzy Measures. Proceedings of the Third Australian and New Zealand Conference on Intelligent Information Systems. 220–225
44. Chaikla, N., Qi, Y.L. (1999): Genetic Algorithms in Feature Selection, IEEE International Conference on Systems, Man, and Cybernetics. 5, 538–540
45. Chao, J., Hoshino, M., Kitamura, T., Masuda, T. (2001): A Multilayer RBF Network and Its Supervised Learning. International Joint Conference on Neural Networks. 3, 1995–2000
46. Chapelle, O., Haffner, P., Vapnik, V.N. (1999): Support Vector Machines for Histogram-Based Image Classification. IEEE Transactions on Neural Networks. 10, 1055–1064
47. Chatfield C. (1989): The Analysis of Time Series, An Introduction. Chapman & Hall, Fourth Edition. Neural York.
48. Chen, S., Cowan, C.F.N., Grant, P.M. (1991): Orthogonal Least Squares Learning Algorithm for Radial Basis Function Networks. IEEE Transactions on Neural Networks. 2, 302–309
49. Chen, S. (1995): Regularised OLS Algorithm with Fast Implementation for Training Multi-Output Radial Basis Function Networks. Fourth International Conference on Artificial Neural Networks. 290–294
50. Chen, K., Yu, X. and Chi, H. (1997): Combining Linear Discriminant Functions with Neural Networks for Supervised Learning. Neural Computing and Applications. 6, 19–41
51. Chen, X., Cheung, S.T., So, S., Fan, S.T., Barry, C. (2002): Gene Expression Patterns in Human Liver Cancers. Molecular Biology of Cell. 13, 1929–1939

52. Chen, C.T., Azimi-Sadjadi, M.R and Yuan, C.H. (1997): Signal Representation Using Adaptive Wavelet-Net. International Conference on Neural Networks. 4, 2225–2230
53. Cline, M., Diekhans, M., Grate, L., Karplus, K., Barrett, C., and Hughey, R.: Predicting protein structure using only sequence information. Proteins: Struct. Funct. Genet. S3: 121–125, 1999.
54. Chothia, C., Brenner, S.E., and Hubbard, T.J.P. (1998): Assessing sequence comparison methods with reliable structurally identified distant evolutionary relationships. Proceedings of the National Academy of Sciences. USA, 95, 6073–6078
55. Chou, P.Y., Fasma, U.D. (1974): Prediction of Protein Conformation. Biochemistry. 13, 211–215
56. Chow, T., Leung, C.T. (1996): Performance enhancement using nonlinear preprocessing. IEEE Transactions on Neural Networks. 7, 1039–1042
57. Chu, F., Wang, L.P. (2003): Gene expression data analysis using support vector machines. Proceedings of the International Joint Conference on Neural Networks. 3, 2268–2271
58. Chu, F., Wang, L.P. (2004): Cancer Classification with Microarray Data using SVMs. Bioinformatics using Computational Intelligence Paradigms, Springer-Verlag, Udo Seiffert and Lakhmi C. Jain (Eds), Sprinter, 167–189, 2005.
59. Chu, F., Xie, W., Wang, L.P. (2004): Gene Selection and Cancer Classification Using a Fuzzy Neural Network, IEEE Annual Meeting of the Fuzzy Information. 2, 555–559.
60. Coggins, J.M. (1999): Non-Linear Feature Space Transformations. IEE Colloquium on Applied Statistical Pattern Recognition (Ref. No. 1999/063), 17/1–17/5
61. Coifman, R.R., Meyer, Y., Wickerhauser, M.V. (1992): Entropy-Based Algorithm for Best Basis Selection. IEEE Transactions on Information Theory. 38(2), 713–718
62. Cortes, C., Vapnik, V.N. (1995): Support Vector Networks. Machine Learning. 20, 273–297
63. Coskun, N., Yildirim, T. (2003): Proceedings of the International Joint Conference on Neural Networks. 2, 1223–1226
64. Costa, G.R.M.D., Langona, K., Alves, D.A. (1998): A new approach to the solution of the optimal power flow problem based on the modified Newton's method associated to an augmented Lagrangian function. 1998 International Conference on Power System Technology. 2, 909–913
65. Cover, T.M. (1965): Geometrical and Statistical Properties of Systems of Linear Inequalities with Applications in Pattern Recognition. IEEE Transactions on Electronic Computers. EC-14, 326–334
66. Craven, M.W., Shavlik, J.W. (1996): Extracting Tree-structured Representations of Trained Networks. In: Touretzky, D., Mozer, M., Hasselmo, M. (eds.), Advances in Neural Information Processing Systems. 8, Cambridge, MIT Press.
67. Crammer, K., Singer, Y. (2001): On the Algorithmic Implementation of Multiclass Kernel-based Vector Machines. Journal of Machine Learning Research. 2, 265–292
68. Cuff, J.A. and Barton, G.J. (1999): Evaluation and improvement of multiple sequence methods for protein secondary structure prediction. Proteins: Struct. Funct. Genet, 34, 508–519
69. Cybenko, G. (1989): Approximation by superposition of a sigmoidal function. Mathematics of Control, Signals, and Systems. 2(4), 303–314

70. Darbari, A. (2000): Rule Extraction from Trained ANN: A Survey. Technical Report WV-2000-03, Knowledge Representation and Reasoning Group, Department of Computer Science, Dresden University of Technology, Dresden, Germany
71. Dash, M., Liu, H., Yao, J. (1997): Dimensionality Reduction of Unsupervised Data. Ninth IEEE International Conference on Tools with Artificial Intelligence. 532–539
72. Davids, D.R. (1964): A Correlation between Amino Acid Composition and Protein Structure. Journal of Molecular Biology. 9, 605–609
73. David, V., Sanchez, A. (1995): Robustization of a Learning Method for RBF Networks. Neuralcomputing. 9, 85–94
74. Denoeux, T., Lengelle, R. (1993): Initializing back propagation network with prototypes. Neural Computation. 6, 351–363
75. Deutsch, J.M. (2003): Evolutionary Algorithms for Finding Optimal Gene Sets in Microarray prediction. Bioinformatics. 19, 45–52
76. Devijver, P.A., Kittler, J. (1982): Pattern Recognition: a Statistical Approach. Prentice-Hall International Inc., London
77. Devore, J.,n Peck, R. (1997): Statistics: the Exploration and Analysis of Data (3rd edition). Duxbury Press, Pacific Grove, CA
78. Dougherty, J., Kohavi, R., Sahami, M. (1995): Supervised and Unsupervised Discretization of Continuous Features. Proceedings of the 12th International Conference on Machine Learning. 194–202
79. Drucker, N., Donghui, W., Vapnik, V.N. (1999): Support vector machines for spam categorization. IEEE Transactions on Neural Networks. 10, 1048–1054
80. Dubois, D., Prade, H. (1982): A Unifying View of Comparison Indices in a Fuzzy Set Theoretic Framework. In: Yager, R.R. (eds.): Fuzzy Sets and Possibility Theory: Recent Developments. Pergamon, New York
81. Duch, W., Adamczak, R., Grabczewski, K. (2001): A new methodology of extraction, optimization and application of crisp and fuzzy logical rules. IEEE Transactions on Neural Networks. 12, 277–306
82. Duda, R.O., Hart, P.E.: Pattern Classification and Scene Analysis. Wiley, New York (1973)
83. Dudoit, S., Fridlyand, J., Speed, T. (2002): Comparison of discrimination methods for the classification of tumors using gene expression data. J Am Stat Assoc. 97, 77–87
84. Duro, R.J., Reyes, J.S. (1999): Discrete-time backpropagation for training synaptic delay-based artificial neural networks. IEEE Transactions on Neural Networks. 10(4), 779–789
85. Dzielinski, A., Zbikowski, R. (1995): Multidimensional sampling aspects of neurocontrol with feedforward networks. Fourth International Conference on Artificial Neural Networks. 240–244
86. Eiben, A.E., Smith, J.E.: Introduction to evolutionary computing. Springer, New York (2003)
87. Eisen, M.B., Brown, P.O. (1999): DNA arrays for analysis of gene expression. Methods Enzymol. 303, 179–205
88. Fahlman, S.E., Lebiere, C. (1990): The cascade-correlation learning architecture. In: Touretzky, D.S. (eds.): Advances in Neural Information Processing Systems II. Morgan Kaufmann, San Mateo, CA. 524–532
89. Fletcher, R. (1987): Practical Methods of Optimization (2nd Edn.). Wiley, New York

90. Fletcher, R., and Reeves, C.M. (1964): Function minimization by conjugate gradients. the Computer Journal. 7, 149–154
91. Fravolini, M.L., Campa, G., Napolitano, K., Song, Y. (2001): Minimal resource allocating networks for aircraft SFDIA. 2001 IEEE/ASME International Conference on Advanced Intelligent Mechatronics. 2, 1251–1256
92. Frawley, W.J., Piatetsky-Shapiro, G., Matheus, C.J. (1991): Knowledge Discovery in Databases: An Overview. In: Piatetsky-Shapiro G., Frawley, W.J. (eds): Knowledge discovery in databases, AAAI Press/MIT Press. 1–27
93. Frayman, Y., Wang, L.P. (1998): Data Mining Using Dynamically Constructed Fuzzy Neural Networks. In: Wu, X., Kotagiri, R., Korb, K.B. (Eds.): Research and Development in Knowledge Discovery and Data Mining. Lecture Notes in Artificial Intelligence, Proceedings of the Second Pacific-Asia Conference on Knowledge Discovery and Data Mining (PAKDD-98). 1394, 122–131
94. Frayman, Y., Ting, K.M., Wang, L.P. (1999): A fuzzy neural network for data mining: dealing with the problem of small disjuncts. International Joint Conference on Neural Networks. 4, 2490–2493.
95. Frayman, Y., Wang, L.P. (1997): Torsional vibration control of tandem cold rolling mill spindles: a fuzzy neural approach, Proc. the Australia-Pacific Forum on Intelligent Processing and Manufacturing of Materials. Eds. T. Chandra, S.R. Leclair, J.A. Meech, B. Verma, M. Smith, B. Balachandran, 1: Intelligent Systems Applications. 89–94
96. Frayman, Y., Wang, L.P. (1997): A fuzzy neural approach to speed control of an elastic two-mass system. Proc. 1997 International Conference on Computational Intelligence and Multimedia Applications (Feb. 10, Gold Coast, Queensland, ISBN 086857 7618), B. Verma and X. Yao (eds.), 341–345
97. Frayman, Y., Wang, L.P. (1998): Data mining using a self-constructed fuzzy neural network. Research and Development in Knowledge Discovery and Data Mining. X. Wu, R. Kotagiri, and K.B. Korb (Eds.), 122–131 (Proc. the Second Pacific-Asia Conference on Knowledge Discovery and Data Mining. 15-17 April 1998, Melbourne)
98. Frayman, Y., Wang, L.P., Wan, C. (2002): Cold rolling mill thickness control using the cascade-correlation neural network. Control and Cybernetics. 31, 327–342
99. Frayman, Y., Wang, L.P. (2002): A Dynamically-constructed fuzzy neural controller for direct model reference adaptive control of multi-input-multi-output nonlinear processes. Soft Computing. 6, 244–253
100. Fu, X.J., Wang, L.P. (2001): Rule extraction by genetic algorithms based on a simplified RBF neural network. Proceedings of the 2001 Congress on Evolutionary Computation. 2, 753–758
101. Fu, X.J., Wang, L.P. (2001): Linguistic rule extraction from a simplified RBF neural network. Computational Statistics (a special issue on Data Mining and Statistics). 16(3), 361–372
102. Fu, X.J., Wang, L.P. (2002): Rule extraction using a novel gradient-based method and data dimensionality reduction. Proceedings of the 2002 International Joint Conference on Neural Networks. 2, 1275–1280
103. Fu, X.J., Wang, L.P., Chua K.S; Chu F. (2002): Training RBF Neural Networks on Unbalanced Data. International Conference on Neural Information Processing (ICONIP'02). 2, 1016–1020

104. Fu, X.J., Wang, L.P. (2002): A GA-based RBF classifier with class-dependent features. Proceedings of the 2002 Congress on Evolutionary Computation. 2, 1890–1894
105. Fu, X.J., Wang, L.P. (2002): Rule extraction using a novel gradient-based method and data dimensionality reduction. Proceedings of the 2002 International Joint Conference on Neural Networks. 2, 1275–1280
106. Fu, X.J., Wang, L.P. (2002): Rule extraction from an RBF classifier based on class-dependent features. Proceedings of the 2002 Congress on Evolutionary Computation. 2, 1916–1921
107. Fu, X.J., Wang, L.P. (2003): Data dimensionality reduction with application to simplifying RBF network structure and improving classification performance. IEEE Transactions on Systems, Man and Cybernetics, Part B. 33, 399–409
108. Fu, X.J., Ong, C.J., Keerthi, S.S., Guang, H.G., Goh, L.P.: Extracting the Knowledge Embedded in Support Vector Machines. International Joint Conference on Neural Networks. 1, 2004.
109. Fu, L.M. (1993): Knowledge-based connectionism for revising domain theories. IEEE Transactions on Systems, Man, Cybernetics. 23, 173–182
110. Fu, L.M. (1999): Knowledge discovery by inductive neural networks. IEEE Transactions on Knowledge and Data Engineering. 11, 992–998
111. Fukumi, M., Akamatsu, N. (1998): Rule extraction from neural networks trained using evolutionary algorithms with deterministic mutation. The 1998 IEEE International Joint Conference on Computational Intelligence. 1, 686–689
112. Fung, G.S.K., Liu, J.N.K., Chan, K.H., Lau, R.W.H. (1997): Fuzzy genetic algorithm approach to feature selection problem. IEEE International Conference on Fuzzy Systems. 1, 441–446
113. Funahashi K. I. (1989): On the approximate realization of continuous mappings by neural networks. Neural Networks. 2(3), 183–192
114. Furey, T.S., Cristianini, N., Duffy N., Bednarski, D.W., Schummer, M., Haussler, D. (2000): Support vector machine classification and validation of cancer tissue samples using microarray expression data. Bioinformatics. 16, 906–914
115. Gaines, B.R. (1996): Transforming Rules and Trees into Comprehensible Knowledge Structures. In: Advances in Knowledge Discovery and Data Mining. MIT Press. 205–229
116. Gallant, S.I. (1998): Connectionist Expert Systems. Communications of the ACM. 31, 152–169
117. Geva, S., Wong, M.T., Orlowski, M. (1997): Rule extraction from trained artificial neural network with functional dependency preprocessing. Proc. First International Conference on Knowledge-Based Intelligent Electronic Systems. 2, 559–564
118. Giles, C.L., Chen D., Sun G., Chen, H., Lee, Y. ,Goudreau, M.W. (1995): Constructive learning of recurrent neural networks: Limitations of recurrent cascade correlation and a simple solution. IEEE Transactions on Neural Networks. 6, 829–836
119. Girosi, F., Mukherjee, S., Osuna, E. (1997): Nonlinear prediction of chaotic time series using support vector machines. Proceedings of the 1997 IEEE Workshop on Neural Networks for Signal Processing. 511–520
120. Golub, G., Kahan, W. (1965): Calculating the singular values and pseudo-inverse of a matrix. Society of Industrial and Applied Mathematics Journal on Numerical Analysis. 205–224

121. Golub, T., Slonim, D.K., Tamayo, P., Huard, C., Gaasenbeek, M., Mesirov, J.P., Coller, H., Loh, M.L., Downing, J.R., Caligiuri, M.A., Bloomfield, C.D., Lander, E.S. (1999): Molecular classification of cancer:class discovery and class prediction by gene expression monitoring. Science. 286, 531–536
122. Gong, X.H., Kuh, A. (1999): Support vector machine for multiuser detection in CDMA communications. Conference Record of the Thirty-Third Asilomar Conference on Signals, Systems, and Computers. 1, 680–684
123. Gonzalez, A., Perez, R. (2001): Selection of relevant features in a fuzzy genetic learning algorithm. IEEE Transactions on Systems, Man and Cybernetics, Part B. 31, 417–425
124. Gribskov, M., McLachlan A.D., Eisenberg D. (1987): Profile analysis: detection of distantly related proteins. Proceedings of the National Academy of Sciences. USA, 84(13), 4355–4358
125. Gupta, M.M., Rao, D.H. (1994): On the principles of fuzzy neural networks. Fuzzy Sets and Systems. 61, 1–18
126. Gupta, A., Sang, P., Lam, S.M. (1999): Generalized Analytic Rule Extraction for feedforward neural networks, IEEE Transactions on Knowledge and Data Engineering. 11, 985–991
127. Hagan, M.T., Menhaj, M.B. (1994): Training Feedforward Networks with the Marquardt Algorithm. IEEE Transactions on Neural Networks. 5, 989–993
128. Halgamuge, S.K., Poechmueller, W., Pfeffermann, A., Schweikert, P., Glesner, M. (1994): A new method for generating fuzzy classification systems using RBF neurons with extended RCE learning Neural Networks. IEEE World Congress on Computational Intelligence. 3, 1589–1594
129. Han, J.W. and Kamber ,M. (2001): Data Mining: Concepts and Techniques. Morgan Kaufmann, 2001
130. Hand, D.J., Mannila, H., and Smyth, P. (2001): Principles of Data Mining. MIT Press, 2001
131. Hartman, E.J., Keele, J.D. and Kowalshi, J.M. (1990): Layered neural networks with gaussian hidden units as universal approximations. Neural Computation. 2(2), 210–215
132. Haykin, S. (1994): Neural Networks: A Comprehensive Foundation. Macmillan College Publishing Company, Inc. NJ.
133. Haykin, S. (1999): Neural Networks: A Comprehensive Foundation. Upper Saddle River, NJ: Prentice Hall, 2nd ed.
134. Haykin, S., Principe, J. (1998): Making sense of a complex world. IEEE Signal Processing Magazine. 15(3), 66–81
135. Higgins, C.M., Goodman, R.M. (1994): Fuzzy rule-based networks for control. IEEE Transactions on Fuzzy Systems. 2, 82–88
136. Higgs, B.: An alternative approach to two-group classification, www.binf.gmu.edu/ jsolka/s2003/csi739/projects/higgs_rep.doc
137. Hipp, J., Guntzer, U., Nakhaeizadeh, G. (2000): Algorithms for Association Rule Mining - A General Survey and Comparison, ACM SIGKDD. 2, 58–64
138. Hirasawa, K. , Matsuoka, T., Ohbayashi, M., Murata, J. (1997): Evaluation of multi-layered RBF networks. 1997 IEEE International Conference on Systems, Man, and Cybernetics. 1, 908–911
139. Hofmann, A., Schmitz, C., Sick, B. (2003): Rule extraction from neural networks for intrusion detection in computer networks, IEEE International Conference on Systems, Man and Cybernetics. 2, 1259–1265

140. Holley, H.L., Karplus, M. (1989): Protein secondary structure prediction with a neural network. Proceedings of the National Academy of Sciences. USA. 86, 152–156
141. Holte, R.C., Acker, L.E. and Porter, B.W. (1989): Concept learning and the problem of small disjuncts. Proc. 11th International Joint Conference on Artificial Intelligence. 813–818
142. Honda, H., Kobayashi, T. (2003): Selection of Causal Gene Sets from Gene Expression Profiles Using GeneFis, New Software Based on FNN. Genome Informatics. 14, 272–273
143. Horikawa, S., Furuhashi, T., Uchikawa, Y. (1992): On fuzzy modeling using fuzzy neural networks with the backpropagation algorithm, IEEE Transactions on Neural Networks. 3, 801–806
144. Hornik, K., Stinchcombe, M., and White, H. (1989): Multilayer feedforward networks are universal approximators. Neural Networks, 2(5), 359–366
145. Hruschka, E.R., Ebecken, N.F.F. (1999): Rule extraction from neural networks: modified RX algorithm", International Joint Conference on Neural Networks. 4, 2504–2508
146. Hruschka, E.R. and Ebecken, N.F.F. (2000): Applying a clustering genetic algorithm for extracting rules from a supervised neural network. International Joint Conference on Neural Networks. 3, 407–412
147. Hsu, C.W., Lin, C.J. (2002): A comparison of methods for multiclass support vector machines. IEEE Transactions on Neural Networks. 13, 415–425
148. Hua, S., Sun, Z. (2001): A novel method of protein secondary structure prediction with high segment overlap measure: support vector machine approach. Journal of Molecular Biology. 308, 397–407
149. Huang, S.H., Endsley, M.R. (1997): Providing understanding of the behavior of feedforward neural networks. IEEE Transactions on Systems, Man, and Cybernetics. 27, 465–474
150. Huber, K.P., Berthold, M.R. (1995): Building precise classifiers with automatic rule extraction. IEEE International Conference on Neural Networks. 3, 1263–1268
151. Ishibuchi, H., Tanaka, H., Okada, H. (1994): Interpolation of fuzzy if-then rules by neural networks. International Journal of Approximate Reasoning. 10, 3–27
152. Ishibuchi, H., Murata, T., Turksen, I.B. (1995): Selecting linguistic classification rules by two-objective genetic algorithms. Proc. IEEE International Conference on Systems, Man, and Cybernetics. 2, 1410–1415
153. Ishibuchi, H., Nozaki, K., Yamamoto, N., Tanaka, H. (1995): Selecting fuzzy if-then rules for classification problems using genetic algorithms. IEEE Transactions on Fuzzy Systems. 3, 260–270
154. Ishibuchi, H., Nii, M. (1996): Generating fuzzy if-then rules from trained neural networks: linguistic analysis of neural networks. IEEE International Conference on Neural Networks. 2, 1133–1138
155. Ishibuchi, H., Nii, M., Murata, T. (1997): Linguistic rule extraction from neural networks and genetic-algorithm-based rule selection. Proc. International Conference on Neural Networks. 4, 2390–2395
156. Ishibuchi, H., Murata, T. (1998): Multi-objective genetic local search for minimizing the number of fuzzy rules for pattern classification problems. Fuzzy Systems Proceedings, IEEE World Congress on Computational Intelligence. 2, 1100–1105

157. Ishibuchi,H.,Sotani,T.,Murata,T. (1999): Tradeoff Between The Performance of Fuzzy Rule-based Classification Systems and The Number of Fuzzy If-then Rules. 18th International Conference of the North American Fuzzy Information Processing Society. 125–129
158. Jain, A., Zongker, D. (1997): Feature Selection: Evaluation, Application, and Small Sample Performance. IEEE Transactions on Pattern Analysis and Machine Intelligence. 19. 153–158
159. Jain, A.K., Murty, M.N., Flynn, P.J. (1999): Data Clustering: A Review. ACM Computing Surveys. 31(3), 264–323
160. Jang, J.S.R., Sun, C.T. (1993): Functional equivalence between radial basis function networks and fuzzy inference systems. IEEE Transactions on Neural Networks. 4, 156–159
161. Jang, J.R. (1996): Input Selection for ANFIS Learning. Proceedings of the Fifth IEEE International Conference on Fuzzy Systems. 2, 1493–1499
162. Jiang, Y., Zhou, Z.H., Chen, Z.Q. (2002): Rule Learning Based on Neural Network Ensemble. Proceedings of the 2002 International Joint Conference on Neural Networks. 2, 1416–1420
163. Joachims, T. (2000): Estimating the Generalization Performance of a SVM Efficiently. Proceedings of the Seventeeth International Conference on Machine Learning (ICML),Morgan Kaufmann
164. Jones, D.T. (1999): Protein secondary structure prediction based on position-specific scoring matrices. Journal of Molecular Biology. 292, 195–202
165. Juang, C., Lin, C. (1999): A recurrent self-organizing neural fuzzy inference network. IEEE Transactions on Neural Networks. 10, 828–845
166. Kambhatla, N., Leen, T.K. (1993): Fast Non-linear Dimension Reduction. IEEE International Conference on Neural Networks. 3, 1213–1218
167. Katayama, R., Kajitani, Y., Kuwata, K., Nishida, Y. (1993): Self Generating Radial Basis Function as Neuro-Fuzzy Model and Its Application to Nonlinear Prediction of Chaotic Time Series. Second IEEE International Conference on Fuzzy Systems. 1, 407–414
168. Kawatani, T., Shimizu, H. (1998): Handwritten Kanji Recognition With The LDA Method. Fourteenth International Conference on Pattern Recognition. 2, 1301–1305
169. Kay, S. (2000): Sufficiency, Classification, and The Class-Specific Feature Theorem. IEEE Transactions on Information Theory. 46, 1654–1658
170. Kaylani, T., Dasgupta, S. (1994): A New Method for Initializing Radial Basis Function Classifiers. IEEE International Conference on systems, Man, and Cybernetics. 3, 2584–2587
171. Keller, J.M., Yager, R.R., Tahani, H. (1992): Neural Network Implementation of Fuzzy Logic. Fuzzy Sets Syst. 45, 1–12
172. Kendrew, J.C., Dickerson, R.E., Strandberg, R.G., Hart, R.G., Davies, D.R., Phillips, D.C., and Shore, V.C.: Structure of Myoglobin (1960): A Three-dimensional Fourier synthesis at 2 Å resolution. Nature. 185, 422–427
173. Khan, J., Wei J.S., Ringner, M., Saal, L.H., Ladanyi, M., Westermann, F., Berthold, F., Schwab, M., Antonescu, C.R., Peterson, C., Meltzer, P.S. (2001): Classification and Diagnostic Prediction of Cancers Using Gene Expression Profiling and Artificial Neural Networks. Nature Medicine. 7, 673–679
174. Khan, E., Unal, F. (1994): Recurrent Fuzzy Logic Using Neural Networks. Proceedings 1994 IEEE Nagoya World Wisepersons Workshop. 48–55

175. Kneller, D.G., Cohen, F.E., Langridge, R. (1990): Improvements in Protein Secondary Structure Prediction by An Enhanced Neural Network. Journal of Molecular Biology. 214, 171–82
176. Knerr, S., Personnaz, L., Dreyfus, G. (1990): Single-layer Learning Revisited: A Stepwise Procedure for Building and Training A Neural Network. Neurocomputing: Algorithms, Architectures and Applications, Springer
177. Kohonen, T. (1990): Improved Versions of Learning Vector Quantization. Proc. IEEE International Joint Conference on Neural Networks. 1, 545–550
178. Koike, T., Lopez, R., Gibson, T.J., Higgins, D.G., Chenna, R., Sugawara, H., and Thompson, J.D. (2003): Multiple sequence alignment with the clustal series of programs. Nucleic Acids Research. 31, 3497–3500
179. Kolen, J.F., Pollack, J.B. (1990): Back Propagation is Sensitive to Initial Conditions. Technical Report. TR 90-JK-BPSIC, Laboratory for Artificial Intelligence Research, Computer and Information Science Department.
180. Kononenko, I. (1994): Estimating Attributes: Analysis and Extension of RELIEF. Proceedings of European Conference on Machine Learning. 171–182
181. Kosko, B. (1992): Neural Networks and Fuzzy Systems. Prentice-Hall: Englewood Cliffs, NJ
182. Kosko, B. (1993): Fuzzy Thinking: The New Science of Fuzzy Logic. Flamingo, London
183. Kubat, M. (1998): Decision Trees Can Initialize Radial-basis-function Networks. IEEE Transactions on Neural Networks. 813–821
184. Kumar, R., Kulkarni, A., Jayaraman, V.K., Kulkarni, B.D. (2004): Symbolization Assisted SVM classifier for Noisy Data. Pattern Recognition Letters. 25, 495–504
185. Kuncheva, L.I., Jain, L.C. (2000): Designing Classifier Fusion Systems by Genetic Algorithms. IEEE Transactions on Evolutionary Computation. 4, 327–336
186. Lapedes, A. and Farber, R. (1987): Nonlinear signal processing using neural network: Prediction and system modeling. Los Alamos Nat. Lab, Technical Report. LA-UR-872662
187. Lazzerini, B., Marcelloni, F. (2002): Feature Selection Based on similarity. Electronics Letters. 38, 121–122
188. Lee, C.C. (1990): Fuzzy Logic in Control Systems: Fuzzy Logic Controller. IEEE Transactions on Systems, Man and Cybernetics. 20, 404–436
189. Lee, J., Beach, C.D., Tepedelenlioglu, N. (1996): Channel Equalization Using Radial Basis Function Network. IEEE International Conference on Acoustics, Speech, and Signals. 3, 1719–1722
190. Lehtokangas, M., Saarinen, J., Kaski, K., Huuhtanen, P. (1995): Initialization weights of a multiplayer perceptron by using the orthogonal least squares algorithm. Neural Computation. 7, 982–999
191. Levenberg K.(1944): A method for the solution of certain problems in least squares. Quarterly of Applied Mathemetics. 2, 164–168
192. Li, L., Weinberg, C.R., Darden, T.A., Pedersen, L.G. (2001): Gene Selection for Sample Classification Based on Gene Expression Data: Study of Sensitivity to Choice of Parameters of The GA/KNN Method. Bioinformatics. 17, 1131–1142
193. Liang, Y. and Page, E.W. (1997): Multiresolution Learning Paradigm and Signal Prediction. IEEE Transactions on Signal Processing. 45(11), 2858–2864
194. Lim, V.I. (1974): Structural principles of the globular organization of protein chains. A stereochemical theory of globular protein secondary structure. Journal of Molecular Biology. 88, 857–872

195. Lin, W.M., Cheng, F.S., Tsay, M.T. (2000): Distribution Feeder Reconfiguration with Refined Genetic Algorithm. IEE Proceedings - Generation, Transmission and Distribution. 147, 349–354
196. Lin, K.M., Lin, C.J. (2003): A Study on Reduced Support Vector Machines. IEEE Transactions on Neural Networks. 12, 1449–1559
197. Liu, P., Li, H. (2004): Efficient Learning Algorithms for Three-Layer Regular Feedforward Fuzzy Neural Networks. IEEE Transactions on Neural Networks. 15, 545–558
198. Liu, C.J., Wechsler, H. (1998): Enhanced Fisher Linear Discriminant Models for Face Recognition. Fourteenth International Conference on Pattern Recognition. 2. 1368–1372
199. Lotlikar, R., Kothari, R. (2000): Bayes-optimality Motivated Linear and Multilayered Perceptron-based Dimensionality Reduction. IEEE Transactions on Neural Networks. 11, 452–463
200. Lu, H.J., Setiono, R., Liu, H. (1996): Effective Data Mining Using Neural Networks. IEEE Transactions on Knowledge and Data Engineering. 8, 957–961
201. Lu, Y., Guo, H., Feldkamp, L. (1998): Robust Neural Learning from Unbalanced Data Samples. IEEE World Congress on computational Intelligence. 3, 1816–1821
202. Ma, X., Salunga, R., Tuggle, J.T., Gaudet, J., Enright, E., McQuary, P., Payette, T., Pistone, M., Stecker, K., Zhang, B.M., Zhou, Y.X., Varnholt, H., Smith, B., Gadd, M., Chatfield, E., Kessler J., Baer, T.M., Erlander, M.G., Sgroi, D.C. (2003): Gene Expression Profiles of Human Breast Cancer Progression. Proceedings of National Academy of Sciences. USA, 100, 5974–5979
203. Maffezzoni, P., Gubian, P. (1994): Approximate Radial Basis Function Neural Networks(RBFNN) to Learn Smooth Relations from Noisy Data. Proceedings of the 37th Midwest Symposium on Circuits and Systems. 1, 553–556
204. Mallat, S.G. (1989): A Theory for Multiresolution Signal Decomposition: The Wavelet Representation. IEEE Transactions on Pattern Recognition. 11(7), 674–693
205. Mallat, S.G. (1989): Multifrequency channel decompositions of images and wavelet models. IEEE Transactions on Acoustics, Speech and Signal Processing [see also IEEE Transactions on Signal Processing]. 37(12), 2091–2110
206. Mamdani, E.H., yAssilian, S. (1995): An Experiment in Linguistic Synthesis with a Fuzzy Logic Controller. International Journal of Man–Machine Studies. 7, 1–13
207. Mao, J.C., Mohiuddin, K., Jain, A.K. (1994): Parsimonious Network Design and Feature Selection Through Node Pruning. Proceedings of the 12th IAPR International Conference on Pattern Recognition. 2, 622–624
208. Marill, T., Green, D.M. (1963): On The Effectiveness of Receptors in Recognition Systems. IEEE transactions on Information Theory. 9, 11–17
209. Matecki, U., Sperschneider, V. (1997): Automated Feature Selection for MLP Networks in SAR Image Classification. Sixth International Conference on Image Processing and Its Applications. 2, 676–679
210. Matthews, B.W.(1975): Comparison of the predicted and observed secondary structure of t4 phage lysozyme. Biochim. Biophys. Acta. 405, 442–451
211. McDonnell, J.R., Waagen, D. (1994): Evolving recurrent Perceptrons for time series modeling. IEEE Transactions on Neural Networks. 5, 24–38

212. McGarry, K.J., yWermter, S., MacIntyre, J. (1999): Knowledge Extraction from Radial Basis Function Networks and Multilayer Perceptrons. Proc. International Joint Conference on Neural Networks. 4, 2494–2497
213. McGarry, K.J., Tait, J., Wermter, S., MacIntyre, J. (1999): Rule-extraction from Radial Basis Function Networks. Proc. Ninth International Conference on Artificial Neural Networks. 2, 613–618
214. McGarry, K.J., MacIntyre, J. (1999): Knowledge Extraction and Insertion from Radial Basis Function Networks. IEE Colloquium on Applied Statistical Pattern Recognition (Ref. No. 1999/063) 15/1–15/6
215. McGrath, R.E. (1996): Understanding Statistics: a Research Perspective. 1st edition, Longman
216. Mezard, M., Nadal, J.P. (1989): Learning in Feedforward Layered Networks: The tiling algorithm. Journal of Physics. 22, 2191–2204
217. Mimura, M., Furukawa, T. (2001): A Recurrent RBF Network-for Nonlinear Channel. IEEE InternationalConference on Acoustics, Speech, and Signal Processing. 2, 1297–1300
218. Mitra, S., Hayashi, Y. (2000): Neuro-Fuzzy Rule Generation: Survey in Soft Computing Framework. IEEE Transactions on Neural Networks. 11, 748–768
219. Moody, J., Darken, C. (1989): Fast Learning in Networks of Locally Tuned Processing Units. Neural Computation. 1, 281–294
220. Muggleton, S., King, R.D., Sternberg, M.J.E. (1992): Protein Secondary Structure Predictions Using Logic-based Machine Learning. Prot Engin. 5, 647–657
221. Murata, J., Itoh, S., Hirasawa, K. (1999): Size-reducing RBF Networks. International Joint Conference on Neural Networks. 2, 1308–1312
222. Murayama, N., Nakamura, E., Okuizumi, H., Sawada, K. (2001): RLGS Profile Segmentation Via a SVM. International Conference on Image Processing. 1, 533–536
223. Murphy, P.M., Aha, D.W., (1994): UCI Repository of machine learning databases [http://www.ics.uci.edu/ mlearn/MLRepository.html]. Irvine, CA: University of California, Department of Information and Computer Science.
224. Nabney, I.T. (1999): Efficient Training of RBF Networks for Classification. Ninth International Conference on Artificial Neural Networks. 1, 210–215
225. Nagano, K. (1977): Triplet Information in Helix Prediction Applied to The Analysis of Super-secondary Structures. Journal of Molecular Biology. 109, 251–274
226. Narazaki, H., Watanabe, T., Yamamoto, M. (1996): Reorganizing Knowledge in Neural Networks: An Explanatory Mechanism for Neural Networks in Data Classification Problem. IEEE Transaction on Systems, Man and Cybernetics, Part B. 26, 107–117
227. Nie, J., Linkens, D. (1995): Fuzzy-Neural Control: Principles, Algorithms and Applications. Prentice-Hall Europe
228. Nielsen, R.H. (1990): Neurocomputing. Addison-Wesley Publishing Company.
229. Nomura, H., Hayashi, I., and Wakami, N. (1992): A learning method of fuzzy inference rules by descent method. First IEEE International Conference on Fuzzy Systems. 203–210
230. Norinder, U. (2003): Support Vector Machine Models in Drug Design: Applications to Drug Transport Processes and QSAR Using Simplex Optimisations and Variable Selection. Neurocomputing. 55, 337–346

231. Nunez, H., Angulo, C., Catala, A. (2002): Rule Extraction from Support Vector Machines. European Symposium on Artificial Neural Networks, Bruges (Belgium), d-side publi. ISBN 2930307-02-1. 107–112
232. Oh, I. , Lee, J., Suen,C.Y. (1998): Using Class Separation for Feature Analysis and Combination of Class-dependent Features. Fourteenth International Conference on Pattern Recognition. 1, 453–455.
233. Oh, I. , Lee, J., Suen, C.Y. (1999): Analysis of Class Separation and Combination of Class-dependent Features for Handwriting Recognition. EEE Transactions on Pattern Analysis and Machine Intelligence. 21(10), 1089–1094
234. Omar, M.K., Hasegawa-Johnson, M. (2003): Strong-sense Class-dependent Features for Statistical Recognition, 2003 IEEE Workshop on Statistical Signal Processing. 490–493
235. Pal, S.K., Mitra, S. (1999): Neuro-Fuzzy Pattern Recognition, Wiley Inter-Science
236. Pan, W.(2002): A Comparative Review of Statistical Methods for Discovering Differentially Expressed Genes in Replicated Microarray Experments,Bioinformatics. 18, 546–554
237. Park, J., Sandberg, I.W. (1993): Approximation and Radial Basis Function Networks. Neural Computation. 5, 305–316
238. Perutz, M.F., Rossmann, M.G., Cullis, A.F., Muirhead, G.,Will, G., North, A.C.T. (1960): Structure of Haemoglobin: A Three-dimensional Fourier Synthesis at 5.5 Resolution. Nature. 185, 416–422
239. Piatetsky-Shapiro, G. (1995): Special Issue on Knowledge Discovery in Databases - from Research to Applications. International Journal of Intelligent Systems. 5(1)
240. Poechmueller, W., Hagamuge, S. K., Glesner, M., Schweikert, P., Pfeffermann, A. (1994): RBF and CBF Neural Network Learning Procedures. 1994 IEEE World Congress on Computational Intelligence. 1, 407–412
241. Poggio, T., Girosi, F. (1990): Networks for Approximation and Learning. Proceedings of the IEEE, 78(9), 1481–1497
242. Polak, E. (1971): Computational Method in Optimization a Unified Approach. Academic Press, New York
243. Pomeroy, S.L., Tamayo, P., Gaasenbeek, M., Sturla, L.M., Angelo, M., McLaughlin, M.E., Kim, J.Y.H., Goumnerova, L.C., Black, P.M., Lau, C., Allen, J.C., Zagzag, D., Olson, J.M., Curran, T., Wetmore, C., Biegel, J.A., Poggio, T., Mukherjee, S., Rifkin, R., Califano, A., Stolovitzky, G., Louis, D.N., Mesirov, J.P., Lander, E.S., and Golub, T.R. (2002): Prediction of Central Nervous System Embryonal Tumor Outcome Based on Gene Expression. Nature. 415, 436–442
244. Pop, E., Hayward, R., Diederich, J. (1994): RULENEG: Extracting Rules from A Trained ANN by Stepwise Negation. Neurocomputing Res. Centre, Queensland University Technol., Brisbane,Qld.,Aust., QUT NRC Technical Report.
245. Powell, M.J.D. (1987): Radial Basis Functions for Multivariable Interpolation: A Review. In J. C. Mason and M. G. Cox (Eds.), Algorithms for Approximation, Oxford: Clarendon Press. 143–167
246. Pudil, P., Ferri, F.J., Novovicova, J., Kittler, J. (1994): Floating Search Methods for Feature Selection with Nonmonotonic Criterion Functions. Pattern Recognition - Conference B: ,Proceedings of the 12th IAPR International Conference on Computer Vision and Image Processing. 2, 279–283

247. Pudil, P., Hovovicova, J. (1998): Novel Methods for Subset Selection with Respect to Problem Knowledge. IEEE Intelligent Systems [see also IEEE Expert]. 13(2), 66–74
248. Qian, N., Sejnowski, T.J. (1988): Predicting the Secondary Structure of Globular Proteins Using Neural Network Models. Journal of Molecular Biology. 202, 865–884
249. Quinlan, J.R. (1986): Induction of Decision Trees. Machine Learning. 6(1), 81–106
250. Quinlan, J.R.(1991): Improved estimates for the accuracy of small disjuncts. Machine Learning. 6(1), 93–98
251. Quinlan, J.R.: C4.5 (1993): Programs for Machine Learning. Morgan Kaufmann: San Mateo, CA
252. Radzik, T. (1992): Newton's Method for Fractional Combinatorial Optimization. 33rd Annual Symposium on Foundations of Computer Science. 659–669
253. Rao, C.R., Mitra, S.K. (1971): Generalized Inverse of Matrices and Its Applications. New York: John Wiley
254. Raymer, M.L., Punch, W.F., Goodman, E.D., Kuhn, L.A., Jain, A. K. (2000): Dimensionality Reduction Using Genetic Algorithms. IEEE Transactions on Evolutionary Computation. 4(2), 164–171
255. Riis, S.K., and Krogh, A. (1995): Improving prediction of protein secondary structure using structured neural networks and multiple sequence alignments. Journal of Computational Biology. 3, 163–183
256. Robson, B., Pain, R.H. (1971): Analysis of the Code Relating Sequence to Conformation in Proteins: Possible Implications for the Mechanism of Formation of Helical Regions. Journal of Molecular Biology. 58, 237–259
257. Rojas, I., Ortega, J., Pelayo, F.J., and Prieto, A. (1999): Statistical analysis of the main parameters in the fuzzy inference process. Fuzzy Sets and Systems. 102, 157–173
258. Rooman, M.J., Kocher, J.P., Wodak, S.J. (1991): Prediction of Protein Backbone Conformation Based on Seven Structure Assignments: Influence of Local Interactions. Journal of Molecular Biology. 221, 961–979
259. Rosenwald, R., et al. (2002): The Use of Molecular Profiling to Predict Survival after Chemotherapy for Diffuse Large-B-cell Lymphoma. The New England Journal of Medicine. 346, 1937–1947
260. Rost, B., Sander, C. (1994): Combining Evolutionary Information and Neural Networks to Predict Protein Secondary Structure. Proteins. 19, 55–72
261. Rost. B (1996): PHD: Predicting One-dimensional Protein Secondary Structure by Profile-based Neural Network. Methods in Enzymology. 266, 525–539
262. Rost, B., Sander, C.(1993): Prediction of protein secondary structure at better than 70% accuracy. Journal of Molecular Biology. 232, 584–599
263. Rost, B., Sander, C., and Schneider, R. (1994): Redefining the goals of protein secondary structure prediction. Journal of Molecular Biology. 235, 13–26
264. Roy, A., Govil, S., Miranda, R. (1995): An Algorithm to Generate Radial Basis Function (RBF)-like Nets for Classification Problems, Neural networks. 8(2), 179–201
265. Roy, A., Govil, S., Miranda, R. (1997): A Neural-network Learning Theory and A Polynomial Time RBF Algorithm. IEEE Transactions on Neural Network. 8(6), 1301–1313

266. Ruggiero, C., Sacile, R., Rauch, G. (1993): A hybrid algorithm for determining protein structure. IEEE Transactions on Biomedical Engineering. 40(11), 1114–1121
267. Ruspini, E.H. (1982): Recent Development in Fuzzy Clustering, Fuzzy Set and Possibility Theory. New York: North Holland. 113–147
268. Ruspini, E.H. (1999): Generation of Qualitative Descriptions of Complex Objects. Proc. 1999 IEEE International Fuzzy Systems Conference (FUZZ-IEEE'99). 1, 222–227
269. Saito, K., Nakano, R. (1998): Medical Diagnostic Expert System Based on PDP Model, IEEE International Conference on Neural Networks. 1, 255–262
270. Saito, T., Takefuji, Y. (1999): Logical Rule Extraction from Data by Maximum Neural Networks. Proceedings of the Second International Conference on Intelligent Processing and Manufacturing of Materials. 2, 723–728
271. Salamov, A., Solovyev, V. (1995): Prediction of Protein Secondary Structure by Combining Nearest-neighbor Algorithms and Multiple Sequence Alignment. Journal of Molecular Biology. 247, 11–15
272. Sasagawa, F., Tajima, K. (1993): Prediction of Protein Secondary Structures by A Neural Network. Computer Applications in the Biosciences. 9, 147–152
273. Sato, M., Tsukimoto, H. (2001): Rule Extraction from Neural Networks via Decision Tree Induction. International Joint Conference on Neural Networks. 3, 1870–1875
274. Schena, M., Shalon, D., Davis, R.W., Brown, P.O. (1995): Quantitative Monitoring of Gene Expression Patterns with A Complementary DNA Microarray. Science. 270, 467–470
275. Scheraga, H.A. (1960): Structural Studies of Ribonuclease III. A model for the Secondary and Tertiary Structure. Journal of the American Chemical Society. 82, 3847–3852
276. Schmitz, G.P.J., Aldrich, C., Gouws, F.S. (1999): ANN-DT: An Algorithm for Extraction of Decision Trees from Artificial Neural Networks. IEEE Transactions on Neural Networks. 109(6), 1392–1401
277. Schneider, R., Sander, C. (1991): Database of homology-derived structures and the structural meaning of sequence alignment. Proteins: Struct. Funct. Genet. 9, 56–68
278. Schneider, R., Sander, C. (1999): Twilight zone of protein sequence alignments. Protein Eng. 12, 85–94
279. Schölkopf, B. (1997): Support Vector Learning
280. Schwenker, F., Kestler, H.A., Palm, G., Hoher, M. (1994): Similarities of LVQ and RBF learning. Proceedings of IEEE International Conference on Systems, Man, and cybernetics. 646–651
281. Schwenker, F., Kestler, H.A., Palm, G. (2000): Radial-basis-function Networks: Learning and Applications. Fourth International Conference on Knowledge-Based Intelligent Engineering Systems and Allied Technologies. 1, 33–43
282. Schwenker, F., Kestler, H.A., Palm, G. (2001): Three Learning Phases for Radial-basis-function Networks, Neural Networks. 14, 439–458
283. Setiono, R., Liu, H. (1996): Symbolic Representation of Neural Networks. Computer. 29, 71–77
284. Setiono, R. (1997): Extracting Rules from Neural Networks by Pruning and Hidden-unit Splitting. Neural Computation. 9(1), 205–225

285. Setiono, R., Leow, W.K. (1999): Generating Rules from Trained Network Using Fast Pruning Neural Networks. International Joint Conference on Neural Networks. 6, 4095–4098
286. Setiono, R. (2000): Extracting M -of-N rules from Trained Neural Networks. IEEE Transactions on Neural Networks. 11(2), 512–519
287. Setiono, R., Leow, L.W. (2000): FERNN: An Algorithm for Fast Extraction of Rules from Neural Networks. Applied Science. 12(1/2), 15–25
288. Setiono, R., Leow, L.W., Zurada, J.M. (2002): Extraction of Rules From Artificial Neural Networks for Nonlinear Regression. IEEE Transactions on Neural Networks. 13(3), 564–577
289. Sharma, S., Ellis, D., Kajarekar, S., Jain, P., Hermansky, H. (2000): Feature Extraction Using Non-linear Transformation for Robust Speech Recognition on the Aurora Database. International Conference on Acoustics, Speech, and Signal Processing. 2, 1117–1120
290. Southern, E.M., Case-Green, S.C., Elder, J.K., Johnson, M., Mir, K.U., Wang, L., Williams, J.C. (1994): Arrays of Complementary Oligonucleotides for Analysing the Hybridisation Behaviour of Nucleic Acids, Nucleic Acids Research. 22, 1368–1373
291. Stephen, F.A.,Warren, G.,Webb, M.,Engene, W.M., David, J.L. (1990): Basic Local Alignment Search Tool. Journal of Molecular Biology. 215, 403–410
292. Stolorz, P., Lapedes, A, Xia, Y. (1992): Predicting Protein Secondary Structure Using Neural Net and Statistical Methods. Journal of Molecular Biology. 225, 363–377
293. Strang, G. and Nguyen, T.(1996): Wavelet and Filter Banks. Wellesley College.
294. Strauss, D.J., Steidl, G. (2002): Hybrid Wavelet-support Vector Classification of Waveforms. J Comput and Appl. 148, 375–400
295. Sugeno, M., Kang, G.T. (1988): Structure Identification of Fuzzy Model, Fuzzy Sets and Systems. 28, 15–23
296. Sugeno, M., Tanaka, K. (1991): Successive Identification of A Fuzzy Model and Its Applications to Prediction of a Complex System, Fuzzy Sets and Systems. 42, 315–334
297. Sun, R. (1999): Knowledge Extraction from Reinforcement Learning, Proc. International Joint Conference on Neural Networks (IJCNN'99). 4, 2554–2559
298. Taha, I.A., Ghosh, J. (1999): Symbolic Interpretation of Artificial Neural Networks. IEEE Transactions on Knowledge and Data Engineering. 11(3), 448–463
299. Takagi, T., Sugeno, M. (1985): Fuzzy Identification of Systems and Its Applications to Modeling and Control. IEEE Transaction Systems, Man and Cybernetics. 15, 116–132
300. Takagi, T., Sugeno, M. (1991): NN-Driven Fuzzy Reasoning. Int. J. Approximate Reasoning. 5, 191–212.
301. Tan, S.C., Lim, C.P. (2004): Application of an Adaptive Neural Network with Symbolic Rule Extraction to Fault Detection and Diagnosis in a Power Generation Plant. IEEE Transactions on Energy Conversion. 19, 369–377
302. Taylor, W.R., Thornton, J.M. (1983): Prediction of Super-secondary Structure in Proteins. Nature. 301, 540–542
303. Teo, K.K., Wang, L.P., Lin, Z. (2001): Wavelet packet multi-layer perceptron for chaotic time series prediction: effects of weight initialization. Proceedings of International Conference of Computational Science ICCS 2001 Part II, San Francisco, CA, May 28-30, 2001, Lecture Notes in Computer Science. 2074 Alexan-

drov, V.N. Dongarra, J.J.; Juliano, B.A.; Renner, R.S.; Tan, C.J.K. (Eds.), 310–317
304. Thomas, J.G., Olsen, J.M., Tapscott, S.J., Zhao, L.P. (2001): An Efficient and Robust Statistical Modeling Approach to Discover Differentially Expressed Genes Using Genomic Expression Profiles. Genome Research. 11, 1227–1236
305. Thrun, S. (1995): Extracting Rules from Artificial Neural Networks with Distributed Representations. Advances in Neural Information Processing Systems. MIT Press, Cambridge, MA. 7
306. Tibshirani, R., Hastie, T., Narasimhan, B., Chu, G.: Class Predicition by Nearest Shrunken Centroids, with Applications to DNA Microarrays. Manuscript is available at http://www-stat.stanford.edu/ tibs/research.html (2002)
307. Tibshirani, R., Hastie, T., Narashiman, B., Chu, G. (2002): Diagnosis of Multiple Cancer Types by Shrunken Centroids of Gene Expression. Proc Natl Acad Sci USA. 99, 6567–6572
308. Tibshirani, R., Hastie, T., Narasimhan, B., Chu, G.(2003): Class Prediction by Nearest Shrunken Centroids with Applications to DNA Microarrays. Statistical Science. 18, 104–117
309. Tickle, A., Andrews, R., Golea, M., Diederich, J. (1998): The Truth Will Come To Light: Directions and Challenges in Extracting the Knowledge Embedded Within Trained Artificial Neural Networks. IEEE Transactions on Neural Networks. 9 1057–1068
310. Ting,K.M. (1994): The problem of small disjuncts: its remedy in decision trees. Proc. 10th Canadian Conf. on Artificial Intelligence. 91–97
311. Ting,K.M. (1994) The problem of atypicality in instance-based learning. The 3rd Pacific Rim Int. Conf. on Artificial Intelligence. 1, 360–366
312. Ting, K.M.(1998): Inducing cost-sensitive trees via instance weighting. Proceedings of The Second European Symposium on Principles of Data Mining and Knowledge Discovery. LNAI-1510, 139–147
313. Towell, G.G., Shavlik, J.W. (1993): Extracting Refined Rules From Knowledge-based Neural Networks. Machine Learning. 13, 71–101
314. Towell, G.G. and Shavlik, J.W. (1994): Knowledge-based Artificial Neural Networks. Artificial Intelligence. 70, 119–165
315. Tsang, E. C.C., Wang, X.Z., Yeung, D.S. (1999): Improving Learning Accuracy of Fuzzy Decision Trees by Hybrid Neural Networks. IEEE International Conference on Systems, Man, and Cybernetics. 3, 337–342
316. Troyanskaya, O., Cantor, M., Sherlock, G., Brown, P., Hastie, T., Tibshirani, R., Botstein, D., Altman, R.B. (2001): Missing Value Estimation Methods for DNA Microarrays. Bioinformatics. 17, 520–525
317. Tsukimoto, H. (2000): Extracting Rules From Trained Neural Networks. IEEE Transactions on Neural Networks. 11, 377–389
318. Tsukimoto, H.: Logical Regression Analysis (2002): From Mathematical Formulas to Linguistic Rules. The Foundation of Data Mining and Knowledge Discovery(FDM02). 59–79
319. Turney, P.D. (1995): Cost-sensitive classification: empirical evaluation of a hybrid genetic decision tree induction algorithm. Artificial Intelligence Research. 2, 369–409
320. Tusher, V.G., Tibshirani, R., Chu, G. (2001): Significant Analysis of Microarrays Applied to the Ionizing Radiation Response. Proceedings of National Academy of Sciences. USA. 98, 5116–5121

321. Umano, M., Okada, T., Hatono, I., Tamura, H. (2000): Extraction of Quantified Fuzzy Rules from Numerical Data. The Ninth IEEE International Conference on Fuzzy Systems. 2, 1062–1067
322. Vafaie, H., De Jong, K. (1992): Genetic Algorithms as a Tool for Feature Selection in Machine Learning. 4th International Conference on Tools with Artificial Intelligence, Arlington
323. Vahed, A., Omlin, C.W. (1999): Rule Extraction from Recurrent Neural Networks Using a Symbolic Machine Learning Algorithm. International Conference on Neural Information Processing. 712–717
324. Van Vuuren, P.A., Hoffman, A.J. (2000): Improved Rule Generation for a Neuro-fuzzy Network. IEEE International Conference on Systems, Man, and Cybernetics. 4, 2845–2850
325. Van, G.T., Suykens, J.A.K., Baestaens, D.E., Lambrechts, A., Lanckriet, G., Vandaele B., De Moor, B., Vandewalle, J. (2001): Financial Time Series Prediction Using Least Squares Support Vector Machines within the Evidence Framework. IEEE Transactions on Neural Networks. 12, 809–821
326. Vapnik, V.N. (1995): The Nature of Statistical Learning Theory. Springer-Verlag, New York
327. Vapnik, V.N. (1998): Statistical Learning Theory. Wiley, New York
328. Vetterli, M., Herley, C. (1992): Wavelets and filter banks: theory and design. IEEE Transactions on Signal Processing. 40(9), 2207–2232
329. Vivarelli, F., Giusti, G., Villani, M., Campanini, R., Fraiselli, P., Compiani, M., Casadio, R. (1995): (1995) LGANN: a Parallel System Combining a Local Genetic Algorithm and Neural Networks for the Prediction of Secondary Structure of Proteins. Computer Application in the Biosciences. 11, 763–769
330. Wang, J., Jean, J. (1993): Resolve Multifont Character Confusion with Neural Network. Pattern Recognition. 26, 173–187
331. Wang, L.P. (1997): On competitive learning. IEEE Transaction on Neural Networks. 8, 1214–1217
332. Wang, L.P., Teo, K.K., Lin, Z. (2001): Predicting time series using wavelet packet neural networks. Proceedings of the 2001 IEEE International Joint Conference on Neural Networks (IJCNN 2001). Washington, DC., USA, 1593–1597
333. Wang, L.P., Fu, X.J. (2005): A simple rule extraction method using a compact RBF neural network. 2nd International Symposium on Neural Networks. China. Lecture Notes in Computer Science, accepted 2005.
334. Wang, L. (1992): Fuzzy Systems Are Universal Approximators. Proc. IEEE International Conference Fuzzy Systems, San Diego
335. Wang, L., Langari, R. (1995): Building Sugeno-type Models Using Fuzzy Discretization and Orthogonal Parameter Estimation Techniques. IEEE Transaction on Fuzzy Systems. 3, 454–458
336. Wang, L.X., Mendel, J.M. (1992): Generating Fuzzy Rules by Learning from Examples. IEEE Transactions on Systems Man, and Cybernetics. 22, 1414–1427
337. Wang, L.X. (1998): Universal Approximation by Hierarchical Fuzzy Systems. Fuzzy Sets and Systems. 93, 223–230
338. Wang, X.Z., Chen, B.H., Yang, S.H., McGreavy, C., Lu, M.L. (1997): Fuzzy Rule Generation From Data for Process Operational Decision Support. Computer and Chemical Engineering. 21, 661–666
339. Wasserman, P.D. (1989): Neural Computing: Theory and Practice. Van Nostrand Reinhold, Co. New York

340. Weigend, A.S., Huberman, B.A., and Rumlhart, D.E. (1992): Predicting Sunspot and Exchange Rates with Connectionist Networks. Nonlinear Modeling and Forecasting. SFI studies in sciences of complexity. XII, 395–432
341. Weiss, G.M. (1995): Learning with rare cases and small disjuncts. Proc. 12th International Conference on Machine Learning. 558–565
342. Weiss,G.M. and Hirsh, H.(1998): The problem with noise and small disjuncts. Proc. 15th International Conference on Machine Learning. 574–578
343. Welch, B.L. (1947): The Generalization of Student's Problem When Several Different Populations Are Involved. Biomethika. 34, 28–35
344. Werbos, P. (1974): Beyond Regression: New Tools for Prediction and Analysis in the Behavioral Sciences. Ph.D. dissertation, Harvard University , Cambridge MA.
345. Whitney, A. (1971): A Direct Method of Nonparametric Measurement Selection. IEEE Transactions on Computation. 20, 1100–1103
346. Witten, I.H. and Frank, E. (2000): Data Mining: Practical machine Learning Tools and Techniques with Java Implementations. Morgan Kaufmann
347. Wu, X. (1995): Knowledge Acquisition from Databases, Ablex Publishing, Norwood
348. Wu, C.H., and Mclarty, J.W. (2000): Neural Networks and Genome Informatics. 1st Edition. Elsevier Health Sciences
349. Xie, W., Wang, L.P. (2003): A Fuzzy Neural Network for System Modeling. Proceedings of Joint 13th International Conference on Artificial Neural Networks and 10th International Conference on Neural Information Processing, Turkey
350. Xing, E.P., Jordan, M.I., Karp, R.M. (2001): Feature Selection for High-dimensional Genomic Microarray Data. Proceedings of the Eighteenth International Conference on Machine Learning. 601–608
351. Yang, J., Honavar, V. (1998): Feature Subset Selection Using a Genetic Algorithm. In: Liu, H., Motoda, H. (eds.): Feature Extraction, Construction and Selection: A Data Mining Perspective. Massachusetts: Kluwer Academic Publishers. 117–136
352. Ye, Q.H., Qin L.X., Forgues, M., He, P., Kim, J.W., Peng, A.C., Simon, R., Li, Y., Robles, A.I., Chen, Y., Ma, Z.C., Wu, Z.Q., Ye, S.L., Liu, Y.K., Tang, Z.Y., Wang, X.W. (2003): Predicting Hepatitis B Virus-positive Metastatic Hepatocellular Carninomas Using Gene Expression Profiling and Supervised Machine Learning. Nature Medicine. 9, 416–423
353. Yuan, H., Tseng, S.S., Wu, G.S., Zhang, F.Y. (1999): A Two-phase Feature Selection Method Using Both Filter and Wrapper. IEEE International Conference on Systems, Man, and Cybernetics. 2, 132–136
354. Zadeh, L.A. (1968): Fuzzy Sets. Information and Control. 8, 338–359
355. Zemla, A., Venclovas, C., Fidelis, K. and Rost, B. (1999): A modified definition of sov, a segment-based measure for protein secondary structure prediction assessment. PROTEINS: Structure, Function, and Genetics. 34, 220–223.
356. Zhang, Y.Q., Fraser, M.D., Gagliano, R.A., Kandel, A. (2000): Granular Neural Networks for Numerical-linguistic Data Fusion and Knowledge Discovery. IEEE Transactions on Neural Networks. 11, 658–667
357. Zhang, Q.; Benveniste, A.(1992): Wavelet networks. IEEE Transactions on Neural Networks. 3(6), 889–898
358. Zhang, X.R. (1994): A hybrid algorithm for determining protein structure. IEEE Expert: Intelligent Systems and Their Applications. 9(4), 66–74

359. Zhao, Q.F. (1997): A Co-evolutionary Algorithm for Neural Network Learning. International Conference on Neural Networks. 1, 432–437
360. Zhao, Q.F. (2001): Evolutionary Design of Neural Network Tree-integration of Decision Tree, Neural Network and GA. Proceedings of the 2001 Congress on Evolutionary Computation. 1, 240–244
361. Zhou, Z.H., Jiang, Y. (2003): Medical Diagnosis with C4.5 Rule Preceded by Artificial Neural Network Ensemble. IEEE Transactions on Information Technology in Biomedicine. 7, 37–42
362. Zhou., Z.H. (2004): Rule Extraction: Using Neural Networks or for Neural Networks? Journal of Computer Science and Technology. 19, 249–253
363. Zupan, B., Bohanec, M., Bratko, I., and Demsar, J.(1997): Machine learning by function decomposition. 14th International Conference on Machine Learning. 421–429

Index